SELECTIONS FROM PLINY'S LETTERS

HANDBOOK

C000162361

CAMBRIDGE LATIN TEXTS

SELECTIONS FROM PLINY'S LETTERS

HANDBOOK

M. B. FISHER and **M. R. GRIFFIN**
Manchester Grammar School

CAMBRIDGE UNIVERSITY PRESS
Cambridge
London - New York - Melbourne

Published by the Syndics of the Cambridge University Press
The Pitt Building, Trumpington Street, Cambridge CB2 1RP
Bentley House, 200 Euston Road, London NW1 2DB
32 East 57th Street, New York, NY 10022, USA
296 Beaconsfield Parade, Middle Park, Melbourne 3206, Australia

First published 1977

Printed in Great Britain by Billing & Son Ltd
Guildford, London and Worcester

Library of Congress Cataloguing in Publication Data

Plinius Caecilius Secundus, C.
Selections from Pliny's letters.

(Cambridge Latin texts)
Bibliography: p.
Includes index.
1. Plinius Caecilius Secundus, C. — Correspondence. I. Fisher, M. B. II.
Griffin, M. R. III. Series.
PA6639.E5F5 1976   876'.01   76-28002
ISBN 0 521 20487 9

CONTENTS

alius aliud: ego beatissimum existimo, qui bonae mansuraeque famae praesumptione perfruitur, certusque posteritatis cum futura gloria vivit. ac mihi nisi praemium aeternitatis ante oculos, pingue illud altumque otium placeat.

(Pliny to Valerius Paulinus: IX. 3*)

FOREWORD

This *Handbook* is the companion to our *Selections from Pliny's Letters* and is intended mainly for use by those who are teaching the Selection as part of the *CLC* 'O' Level course. It is hoped that it may also be of use to students themselves at universities or in sixth forms.

The Introduction hopes to give a compact summary of Pliny's career against the background of his times, as well as some discussion of the Letters. After that, we have dealt with each letter individually on a fixed pattern: in a section called 'The letter' the contents are summarised, the recipient of the letter is introduced and references are given to other letters on the same theme. The next section, 'Notes', is similar to a traditional commentary. The section 'Exploration' is intended to stimulate study of the letter in depth; and the 'Questions' at the end may help such study further. The other parts of the *Handbook*, Bibliography, Appendix, etc., are self-explanatory. Immediately before the letters from Book X there is a biographical note on the Emperor Trajan.

Numerous people have helped us in our writing of this *Handbook*. Our special thanks go to our colleague Robin Griffin, who has guided us to a considerable rethinking of the *Handbook* by his many constructive criticisms. Other help of a similar kind has been given by William Shepherd of the Cambridge University Press, Norman Marlow of Manchester University, Michael Winterbottom of Worcester College, Oxford and David Morton and his colleagues at the CSCP office: to all these people we extend our thanks. We are also grateful to our colleagues in the Classics Department of the Manchester Grammar School for their constructive interest; nor must we forget those to whom we have taught these letters. Our thanks go also to Miss Valerie Marr for so patiently and helpfully typing and retyping the various drafts of this *Handbook*. We of course are responsible for any errors which this book may contain.

Much of this *Handbook* has been written at St Deiniol's Library, Hawarden. We would like therefore to thank the Warden and all others there for the facilities and atmosphere which have helped us so much.

Manchester
April 1976

ABBREVIATIONS

CAH	*Cambridge Ancient History*
Carcopino	
	Carcopino, J., *Daily life in ancient Rome*
CIL	*Corpus Inscriptionum Latinarum*
CLC	*Cambridge Latin Course*
EJ	*Documents illustrating the reigns of Augustus and Tiberius* collected by Victor Ehrenberg and A. H. M. Jones
ILS	*Inscriptiones Latinae Selectae* ed. H. Dessau
JRS	*Journal of Roman Studies*
MW	*Select documents of the principates of the Flavian Emperors including the year of revolution, AD 68-96* ed. M. McCrum and A. G. Woodhead
OCD	*Oxford Classical Dictionary*
Paoli	Paoli, U. E., *Rome, its people, life and customs*
RIB	*The Roman Inscriptions of Britain* ed. R. G. Collingwood and R. P. Wright
S-W	Sherwin-White, A. N., *The Letters of Pliny: a historical and social commentary*
Smallwood (1)	
	Documents illustrating the principates of Gaius, Claudius and Nero ed. E. Mary Smallwood
Smallwood (2)	
	Documents illustrating the principates of Nerva, Trajan and Hadrian ed. E. Mary Smallwood
Syme, *Tacitus*	
	Syme, Sir R., *Tacitus* (two volumes, listed I and II for convenience of reference)

SYMBOLS

Letters referred to thus (I. 6) are letters included in the Cambridge Latin Texts: *Selections from Pliny's Letters*, ed. M. B. Fisher and M. R. Griffin.

Letters referred to thus (II. 1*) are not included in the above Selection.

INTRODUCTION

I Background to Pliny's career

In the second of his letters to the historian Tacitus about the eruption of Vesuvius in August AD 79 Pliny says that he was seventeen at the time of the events he was describing (VI. 20*). This means that he was born late in AD 61 or during the first part of AD 62. His death was perhaps in AD 113.[1] The fifty or so years of his life thus cover the second half of the Emperor Nero's reign, the short reigns of Galba, Otho and Vitellius, the reigns of the three Flavian emperors, Vespasian, Titus and Domitian, that of Nerva and two-thirds of that of Trajan. Pliny's public career was under the final four of these emperors, Titus, Domitian, Nerva and Trajan.

The various parts of Pliny's career formed a pattern more familiar to his contemporaries than to us. It therefore seems reasonable to attempt now to put Pliny's career into its historical context, both by way of introduction to the next section (which contains the often unadorned details of his career) and also as an aid to a better appreciation of Pliny's Letters. With these aims in mind, it will first be necessary to take a look at the political conditions that existed in Rome during Pliny's lifetime and before, and then to examine the kind of political careers open to Pliny and his contemporaries, and then to have a brief look at the law as part of a Roman's career.

Political conditions

Rome had been ruled by emperors for nearly ninety years by the time Pliny was born. Modern historians have often made the start of the principate (the period of rule by emperors) the year 27 BC; in that year Octavian had received the title Augustus and by certain constitutional adjustments he and the Roman senate had ostensibly come to share the ruling of the empire.[2] Certainly the year 27 BC had been an important

---

1 The date of Pliny's appointment to Bithynia, and so his death there less than two years later, is a matter of some controversy. See S-W, 80-2; Syme, *Tacitus*, I. 81.

2 See A. H. M. Jones, 'The Imperium of Augustus', *JRS* XLI, (1951), 112-19 (reprinted in Jones' *Studies in Roman government and law*) for the various constitutional settlements of Augustus' principate.

date. But for the Roman historian Tacitus at least, the defeat by
Octavian of his rivals Mark Antony and Cleopatra four years earlier at
Actium (31 BC) had marked the decisive end of the old system of 'free'
government (the Republic, characterised by political *libertas*) and the
start of autocracy of the emperors. Wistfully recalling the freedom of
opinion possible under the Republic, he referred to the Battle of
Actium and commented that 'the interests of peace demanded the con-
centration of power in the hands of one man' (*Histories*, I. 1). He was
to return to the same theme at the start of a later work, the *Annals*.
There he mentions the *libertas* of the Republic and speaks of Augustus
'who with the name *princeps* received under his control the whole state
wearied by civil discords' (*Annals*, I. 1).

Tacitus regretted the passing of the Republic; but for the great mass
of people in the empire the principate probably brought positive advantages. For the provincials (those living in Roman territories outside
Italy) republican government had often been a tale of criminal misrule
and exploitation. Read Cicero's speech against Verres (70 BC) to see
how a governor could exploit the provincials. The emperors before and
in Pliny's and Tacitus' day had a genuine concern for good government
in the provinces. It is only necessary to read the correspondence that
passed between Pliny in Bithynia and Trajan to see how seriously Pliny
took his work in Bithynia and how evidently concerned the Emperor
himself was.

Then there were the people of Rome. For the mass of the people
there was probably immense relief when Augustus assumed sole charge
of the government.[3] He offered stability after the uncertainties of civil
war before Actium; and even Tacitus admits that 'he won everyone over
with the sweet attractions of peace' (*Annals*, I. 2). Autocracy (for that
was what the principate was or became) meant little difference to the
ordinary people. They had always been clients: and, as under the Republic the great senatorial families had been their patrons, it mattered
little to them now it was the emperor who was effectively their patron.

The *equites* must also have come to welcome the principate. Augustus
gave them right from the start the chance of power and a place in the
administrative hierarchy of the empire (see p. 4). Not every *eques*
would want such power; but for those who did, there must have been a
feeling of enhanced importance, while the settled conditions of the
principate must have been of great benefit to the commercial interests
of the *equites*.

It was in fact the senatorial class that gained least from the principate. Under the Republic the senators had been the ruling class. True,
in the decades before Actium the civil wars had more and more transferred the real power from the senate as a whole to a few great per-

3 See A. H. M. Jones, *op. cit.*, pp. 116-17.

sonalities (men like Sulla, Pompey, Caesar, Mark Antony and Octavian). But Tacitus is probably right: the real break came with the Battle of Actium. Augustus might remark in his *Res Gestae*: 'I transferred the republic from my power to the will of the senate and Roman people' (34). But he was speaking with less attention to honesty than to propaganda. His was a *respublica restituta* but not a *respublica reddita*;[4] he preserved the old republican offices and forms of government (see below, pp. 4-5); but he kept the real power in his own hands. As Tacitus says of Augustus: 'Little by little he began to increase in power. He drew to himself the functions of the senate, of the magistrates and of the laws' (*Annals*, I. 2). Under the emperors the senatorial class ceased to be the rulers and became (with the *equites*) the administrators.

With the power now transferred to an autocrat, it mattered very much to senators in particular what kind of man the emperor was. Pliny's own lifetime provides striking contrasts in emperors. The tradition presents Nero, Otho, Vitellius and Domitian in anything but a favourable light. Galba was, in Tacitus' epigram, 'capax imperii, nisi imperasset' (*Histories*, I. 49), whereas Vespasian — unlike all his predecessors — was changed for the better by being emperor (*Histories*, I. 50). Whatever Tacitus' personal reasons for remembering Domitian with distaste, the senate as a whole found much to dislike in him, and orders were given for Domitian's name to be erased from inscriptions. Trajan won himself a more glowing memory. Pliny and Tacitus were warm in their praises of him (*Panegyric, passim; Agricola,* 3; *Histories,* I. 1; etc.); but, since they were writing during his reign, these praises should be treated with caution. More significantly perhaps, Trajan won praise from those who lived after him. In the middle of the second century AD, Fronto wrote: 'For my part I leave it an open question whether Trajan should be reckoned more distinguished in war than in peace' (*Preamble to the History of Parthian War,* 17). Again, in the fourth century the senate looked for an emperor who would be 'felicior Augusto, melior Traiano' (Eutropius, VIII. 5). With Trajan we are more fortunate than with many emperors: we have his correspondence with Pliny in Bithynia; and from this, and indeed from the general impression of the age gained from Pliny's Letters, we may be able to make up our own minds about the kind of ruler Trajan was and what it may have been like to have lived under the principate, at least in good times.

Political career

Pliny himself was not born into the senatorial class: he could easily have pursued an equestrian career. In fact, his uncle, the Elder Pliny, was an

---

4 Fergus Millar, 'Triumvirate and principate', *JRS* LXIII, (1973), 50-67.

eques, and at the time of his death in August AD 79 he held an equestrian appointment, as Prefect of the Fleet at Misenum (VI. 16). To become an *eques* a man had to be a Roman citizen of two generations' free descent and he also needed sufficient money. This financial qualification had by Pliny's time long been a comparatively modest 400,000 sesterces.[5] The *equites* were originally those who were able to go to war on horseback, like the mediaeval knights. But they became the business class, growing in number as Rome's empire and trade expanded. (The senatorial class were forbidden to trade by law.) Inevitably some of the *equites* would in the end want political power. In 123/122 BC they gained control of the court which tried ex-governors (who were senators) for extortion in their provinces. This may have awakened or partially satisfied their political hopes. But it was Augustus who gave them well-defined posts of responsibility in the administration. These expanded rapidly from a mere twenty-five attested under Augustus[6] and included the prefecture (governorship) of Egypt and the prefecture of Rome's corn supply, as well as the control of certain provincial areas like Judaea. (Pontius Pilate was an *eques.*) Other less ambitious equestrian posts included financial appointments in the provinces (procuratorships) and army appointments, like being the senior centurion of a legion. Some equestrians became very powerful. The Prefects of the Praetorians, because of their closeness to the emperors, could become very influential, like Burrus who was with Seneca effective head of Rome's government for the first eight years of Nero's reign. The Prefects of Egypt were in reality quite as powerful as most senatorial governors. By Trajan's time all great secretariats were held by *equites.* The only thing that a powerful *eques* lacked was the superior status *(dignitas)* of a senator. This was so important to a Roman that *equites* perpetually strove to make their families senatorial. Such ambition may have made the eighteen-year-old Pliny decide not to remain an equestrian like his uncle but follow a senatorial career.

What kind of senatorial career could Pliny at the age of eighteen look forward to? Augustus, in making the senators administrators rather than rulers, had been careful to preserve the traditional republican forms of government. Under the Republic, men of the senatorial class had worked their way up the *cursus honorum* through offices such as the quaestorship and praetorship to the consulship (for the few who reached the top). Augustus studiously preserved this pyramid of promotion for senators. There were still quaestors, tribunes of the people,

---

5 Modern monetary equivalents are deceptive: the distribution of wealth was so much more uneven then than now; but 400,000 sesterces is perhaps something like £100,000. 40,000 sesterces would apparently provide a centurion with his outfit and equipment: VI. 25. A legionary soldier in Domitian's reign was paid 1,200 sesterces per annum.
6 Fergus Millar, *The Roman Empire and its neighbours,* 55ff., for instance.

praetors and even consuls, and Pliny's Letters show that men continued to be anxious to climb the *cursus honorum* (see, for instance, II. 9, where Pliny canvasses for Sextus Erucius as he stands for tribune of the people). The consuls were still *de iure* heads of state; and to allow more of the senators to hold the consulship it became the custom under the emperors for the first two consuls of the year (*consules ordinarii*) to resign after several months and to be replaced by two *consules suffecti* who in their turn might be replaced by two further *consules suffecti*, and so on. Those who had been consuls and praetors were, as under the Republic, chosen for the governorship of provinces. The office of emperor, at first at least, did not constitutionally exist: the emperor was merely *princeps* (first citizen).

But, whatever the constitutional fictions, right from the start the emperors held the real power. They saw to it that they had the men of their choice in the magistracies and other government appointments. Pliny's own career must exemplify this. He had a speedy ride up the *cursus honorum,* reaching the consulship at an early age, before he was forty. (Cornutus Tertullus, his consular colleague in AD 100, who had also not held this magistracy before, was some fifteen to twenty years older than him.) Pliny's success was due in part to insistent political backing from men like Verginius Rufus, Julius Frontinus and Corellius Rufus. But he must also have been assisted by imperial backing — his uncle had been a close friend of Vespasian and he himself was the emperor's quaestor. Tacitus' father-in-law Agricola was consul at about the age of thirty-seven and then governor of Britain for a longer time than any of his predecessors. He could have expected another governorship after his return to Rome, and Tacitus may be right in suspecting a certain jealousy in Domitian towards Agricola. As for Sextus Erucius, Trajan may not have been much bothered about his career: he was standing for *tribunus plebis* about AD 100 but did not become consul till AD 117. When his uncle was dismissed by Hadrian from his post as Prefect of the Praetorians some four years later, Sextus also seems to have gone into the political wilderness, only to be rescued a quarter of a century later, when he was old, for a second consulship and the Prefecture of the City (AD 146) (see II. 9).

There were two types of senatorial career. Men who had an aptitude for soldiering would follow careers that were a combination of civilian and military appointments. Agricola is the classic case of such a career: he was in succession tribune of the soldiers (*circa* AD 61), quaestor (AD 64), tribune of the people (AD 66), praetor (AD 68), *legatus legionis* (AD 71-3), governor of Aquitania (AD 74-7), consul and then governor of Britain (AD 77/8-83/4). Those with little liking for the military life led careers that were exclusively or almost exclusively civilian. So Pliny, after six months as *tribunus militum* in Syria very early in his career, held no further military appointments but still, like

Agricola, worked his way progressively through the quaestorship, tribunate of the people, praetorship and consulship. He and Agricola were both men with successful public lives, following the two senatorial career-patterns encouraged under the emperors.

What Pliny and Agricola and all other senatorials had in common was that they were expected to pursue active careers in the government service. To be born into the senatorial class or to be allowed to join it (as Pliny did) carried with it an obligation to lead a public life. It was up to the emperor to further a senatorial's career, just to let it run its course unaided or positively to thwart it. But a senatorial was not expected to opt out: and it is against this background of obligation to service in public life that Pliny's career and his attitudes must be considered.

Law

Pliny's Letters make absolutely clear his deep involvement in his legal career. The law had always played an important part in the lives of senatorial Romans. With that amateurism that characterised so much of the running of republican Rome, the members of the ruling class had often attended to the law side by side with governing their country and looking after their estates. Those who had a knowledge of the law came originally from this class. They framed the laws, acted as prosecution or defence counsel, representing their clients and even the provincials in court. When we think of the great republican lawyers, Cicero is probably the first name that comes to mind. While his contemporaries made reputations for themselves by their military exploits, he was pursuing a successful career in law at the same time as he was making his name politically. In a similar way Pliny left it to others like Agricola to follow 'mixed' political and military careers, while he followed the law as part of his career. His education will have included training in oratory. At the age of eighteen he was already practising in court, and he went on to make a considerable name for himself in the Centumviral Court.

The material rewards for a barrister were, for a man as financially secure as Pliny, trivial. A maximum of 10,000 sesterces had for some years been the accepted 'fee' due to a barrister appearing in a case; and Juvenal was to remark that there was no money in the profession in Rome (*Satires*, VII. 106ff.). Pliny even implies that he charged (or was expected to charge) no fee, when he appeared in court (VI. 23). There might even be times when a senator would have to give up his legal practice temporarily because of governmental commitments. Thus Pliny ceased to practise while he was Prefect of Saturn, resuming after his consulship (X. 3a*).

It would not be financial reward that called men to the Bar. Barristers could make a name for themselves, something that also brought

with it the possibility of advancement in a man's political career
(IV. 24*). However, there was also a sense of pride in the profession.
Pliny had nothing but praise for young lawyers like Ummidius
Quadratus, a man whom he had evidently helped in his training for
the law and who went on to show promise in the courts (VI. 11*;
VII. 24). The implication is that he held the practice of law in high
regard. One feels that this is one more instance of the conservatism of
the Roman mind. The law had always been in the hands of the upper
class, and men like Pliny would consider it right that it should stay
there.

II Pliny's career

There is little difficulty in obtaining a satisfactorily detailed picture of
Pliny's life. Among modern summaries of it several particularly deserve
mention here: first and foremost, that in Sherwin-White's introduction
to his commentary (pp. 72ff.); then 'The career of Pliny', chapter VII
(pp. 75-85) of Syme's *Tacitus*, vol. I; and Mrs Radice's introduction to
her Penguin Classics translation of Pliny's Letters. However, it may be
helpful also to summarise Pliny's career here as a ready point of reference for those teaching or reading the Selection of Letters, which
this *Handbook* accompanies.

Pliny was born either in AD 61 or 62. He died just over fifty years
later, perhaps in AD 113. He came from Comum (modern Como) in
northern Italy; and throughout his life he retained connections with
and affection for this area, his *patria*. His father died early in Pliny's life.
Verginius Rufus (whose important career included his suppression of
Vindex's rebellion in AD 68 and his refusal to make a bid for the principate) became the young Pliny's guardian. He was evidently an important
influence on Pliny, who wrote of him at the time of his death towards
the end of the century: 'I had many reasons to love him; we came from
the same district and neighbouring towns, and our lands and property
adjoined each other; and then he was left by will as my guardian, and
gave me a father's affection' (II. 1*). Pliny's education included a
period in Rome, where he attended the lectures of the rhetorician
Quintilian and the now lesser known Nicetes Sacerdos, a teacher of
rhetoric from Smyrna.

In August AD 79 the seventeen-year-old Pliny and his mother were
staying at Misenum on the Bay of Naples with Pliny's uncle. When
during the course of that month the volcano Vesuvius erupted, the
uncle — a geographer and scholar — put out to sea to take a closer
look at the eruption as well as to help rescue some friends. In a long
letter to Tacitus (VI. 16) Pliny describes how this expedition led to his
uncle's death. By the terms of his uncle's will Pliny became his heir,
inheriting his estate and so adopting his name to become Gaius Plinius

Luci filius Caecilius Secundus. (To distinguish between the two, the nephew has long been called the Younger Pliny and his uncle the Elder Pliny.) This inheritance, together with bequests from his father, made Pliny a rich man. Pliny's Letters give the clear impression that he never had to worry for money during his adult life.

With sufficient wealth and with the support of influential friends like Verginius Rufus, the youthful Pliny decided to go for a senatorial career. The outlines of the political side of Pliny's career are preserved on an inscription now at Milan (see Appendix no. 1). On this inscription the career is given, not untypically, in reverse order. It shows Pliny starting off with a minor magistracy (he became one of the *decemviri stlitibus iudicandis*, a commission for judging cases involving liberty or citizenship). Then he became a military tribune in the III. Gallica legion in Syria. One wonders how much real soldiering Pliny did. He tells his friend and political associate, Cornutus Tertullus, that he was ordered 'by the consular *legatus* to audit the accounts of the cavalry and infantry divisions' (VII. 31*). His ability with figures was made use of throughout his career, and Pliny became a recognised financial expert. Meanwhile, after his military service Pliny presently became tribune of the people, a quaestor on the emperor's staff, praetor, *praefectus* of the Military Treasury and then *praefectus* of the Treasury of Saturn, before becoming *consul suffectus* in AD 100 (with Cornutus Tertullus as his colleague). This quick sequence of appointments (culminating in the consulship before he was forty), reveals a capacity to survive and do well under such different emperors as Domitian, Nerva and Trajan. Subsequently making somewhat slower progress, he was in charge of the Tiber and its banks and Rome's sewage, from AD 104 for three years, before being sent out to Bithynia by Trajan as his envoy, probably in AD 111.[7]

The Bithynian appointment was, if not the most prestigious, the most important of Pliny's career. Bithynia had particular problems; and Pliny was specially chosen to deal with these problems (see below, pp. 13-16). He was in Bithynia a little less than two years and almost certainly died there (S-W, 82). Perhaps the strain of the post undermined his health. At any rate, the province continued to be a problem and Cornutus Tertullus, Pliny's consular colleague in AD 100, was like Pliny specially appointed to the province (see Appendix no. 2).

The Milan inscription speaks only of Pliny's political career. But Pliny was also a lawyer of distinction. From right at the start of his legal career he spoke frequently in the Court of the Centumviri.[8] He proudly tells Suetonius, the author of those pocket-biographies, *The Twelve Caesars*, of an occasion when he had spoken there when he was

7 For discussions of this controversial date, S-W, 80-2; Syme, *Tacitus*, I. 81.
8 For the Court of the Centumviri, see notes on IV. 16.

young and inexperienced: 'I won my case, and it was that speech which drew attention to me and opened up to me the door of fame' (I. 18*). He told his third wife's grandfather, Calpurnius Fabatus, that in the Court of the Centumviri he was 'in mea arena'(VI. 12*). In IV. 16 he recollected speaking for seven hours on the go in the Centumviral Court, an effort rewarded for him by a young patrician staying the length of the speech to listen to him. He wrote with satisfaction of the congratulations he would often receive in this court (IX. 23*). He also acted in a number of trials of men returning from the governorship of provinces. He appeared for the prosecution against an ex-governor of Baetica, Baebius Massa, in AD 93 and with Tacitus against Marius Priscus, an ex-governor of Africa (on this occasion the Emperor presided — it was evidently a very important case — and Priscus was convicted in January AD 100) (II. 11*). Then twice during the next decade he defended ex-governors of Bithynia. So as a lawyer too Pliny reached the top of his profession.

The law-courts and the official government appointments were the public sides of Pliny's career. But there was also his writing. From an early age he developed literary interests. At fourteen he wrote a Greek tragedy; and on his way back from military service he tried his hand again at verse (VII. 4*). At the time of the eruption of Vesuvius he preferred to stay behind and continue his studies rather than accompany his uncle on his fatal inspection of the eruption (VI. 16). Later, as the effects of the eruption developed, Pliny read from a volume of Livy (VI. 20*). Literature was obviously important to him and he was clearly widely read. He quotes Thucydides; he admires Cicero; he recognises the literary gifts of Tacitus (and tells him so in VII. 20*). While others went to the Circus Maximus, he preferred to spend time among his writing-tablets and notebooks (IX. 6). He loitered with his writing-materials by the nets at a boar-hunt (I. 6) and he escaped whenever he could to his villa at Laurentum, where he could relax and write (I. 9).

Pliny's literary reputation now rests on the ten books of his *Letters*, the first nine of which were revised for publication, on his *Panegyric* to Trajan and on the few snippets of his poetry that he quotes in his *Letters*. In fact, of course, he wrote more than this. Three letters (I. 8*, IX. 14*, IX. 28*) indicate that he revised speeches for publication. He wrote a book of hendecasyllabic verse (IV. 14*) and a volume of poetry in different metres which it took two days to read to a select audience of friends (VIII. 21*). Whatever the friends really thought of this marathon, Pliny's contemporaries apparently had a good opinion of his literary achievements. Some of the praise (reported as it is by Pliny himself) needs treating carefully. Pliny moved in literary circles: he was a friend of Tacitus and Suetonius and other less well-known figures; and literary groups are prone to orgies of mutual congratulation and admiration, as Pliny has not been the last to show. But there is an engaging little story that came to Pliny from Tacitus (IX. 23*). Tacitus had been sitting at the Circus next to a Roman knight. Conversation had turned on several learned topics; and then the knight, not knowing that it was Tacitus to whom he was talking, asked if he was an Italian or provincial. '"You know me", Tacitus replied, "from your reading". To which the man answered, "Then are you Tacitus or Pliny?"' So Pliny was a well-known writer in his own day. In literature, as in the public life of politics and the law, Pliny had succeeded.

III Pliny's 'Letters'

(i) Books I-IX

Pliny's *Letters* divides naturally into two sections: Books I-IX, which are letters to friends; and the largely official correspondence of Book X. Altogether there are almost two hundred and fifty letters in the first nine Books. The people to whom Pliny wrote these letters number over a hundred, and they had entered Pliny's life for many different reasons. There is his third wife Calpurnia (VI. 4*; VI. 7; VII. 5); her grandfather (IV. 1; VII. 16, *inter alias*); literary friends such as Suetonius (I. 18*, etc.) and Tacitus (I. 6; VI. 16; IX. 23*, *inter alias*); friends back in his home-town of Comum, men like Caninius Rufus (I. 3*; II. 8; IX. 33, for instance); professional friends and colleagues, like Cornutus Tertullus (VII. 21; VII. 31*); and so on. Perhaps the number and variety of people to whom Pliny wrote is as good an indication as any that he led an active public and private life.

The letters of Books I-IX were specially revised for publication by Pliny. Consequently the dating of the individual letters, as well as the various dates of publication, presents considerable difficulties in the very first letter in the published collection (I. 1*) Pliny writes to Sextus Erucius' uncle, Septicius Clarus: 'You have often urged me to collect and publish any letters of mine which were composed with some care. I have now made a collection, not keeping to the original order . . . but taking them as they came to hand.' So obviously the letters were not arranged for publication in the order in which they were written. Is Pliny even being honest, when he claims that he arranges them in the order in which they come to hand? It seems best to say no more here than that the nine Books were published (probably not all at once) before Pliny went out to Bithynia. Those who wish to take the question of chronology and publication dates for the letters further should refer to Sherwin-White and Syme, *Tacitus*, II, Appendix 21 (660-4).

The letters themselves cover a whole range of topics of interest to Pliny. To instance one, they are particularly informative about patronage (for a discussion of which, see pp. 20-1). The museums are full of inscriptions which give the outlines of careers of ancient Romans. But what Pliny's letters do is to emphasise the importance, as well as the operation, of patronage on a Roman's career. Pliny himself seems to have been good at giving people a helping hand (even if convention required him as a rich man and a patron to do just this). In I. 19 he offers Romatius Firmus 300,000 sesterces so that he can qualify financially to become an *eques*. Elsewhere Pliny agrees to speak in court if Cremutius Ruso, a young man whom he wants to help in his career, is allowed to act with him (VI. 23). Then there is the relationship between communities and their patrons. Again, any number of inscriptions record public works constructed by men in the towns of which they were patron. But Pliny's letter to his wife's grandfather, explaining that he and his wife must delay on their journey to him in order to be present at the dedication of a temple at Tifernum-on-Tiber, where Pliny was patron, and to attend a celebratory banquet (IV. 1), breathes life into Roman patronage and how it worked.

Aspects of patronage find their way into many of Pliny's letters. Pliny attends a dinner-party and tells a fellow-guest that he is horrified at the way the guests are divided into three groups, each getting food and wine according to its social standing (II. 6). (Pliny himself was in the top group, as he does not fail to point out; so he did not write as a result of personal humiliation). Graded dinner parties were perhaps the rule; but Pliny's attitude to freedmen (V. 19) and slaves (VIII. 16) was enlightened for the time, going beyond what was expected of him as a patron and a master.

There are also a number of letters which are, in part at least, obituary notices of contemporaries (though they differ from the modern newspaper obituary). No selection from Pliny would be complete without one or two of them. The present Selection includes the death of Fannius (V. 5); the pen-sketch of the elderly Ummidia Quadratilla (VII. 24); and the account of the death of Pliny's uncle (VI. 16). The complete set of letters contains letters about Verginius Rufus (II. 1*) and Corellius Rufus (I. 12*) and one concerning the Elder Arria, whose bravery was a by-word (III. 16*).

It is not possible to put all Pliny's letters into social and historical categories. Half the charm of many of them is the way a single letter suddenly throws light on some aspect of life or some incident from Pliny's time. One such letter is that about the Games at Vienna (modern Vienne) in Gaul (IV. 22). Here Pliny strays from his normal practice of a single letter, a single theme, and introduces his readers to a *consilium* of the emperor and an imperial dinner party, as well as to notable men of the day. This letter is surely a must in any Pliny selection (although because of its many historical and social references it is a difficult one from the teaching point of view). But in its way every single one of Pliny's letters helps to build up, however slightly, a more complete picture of the age.

And what kind of portrait of the age do the letters of Pliny present? There is an atmosphere of wealthy calm and refinement. The impression of wealth should be expected. Of the hundred million and more people living in the Roman empire in Pliny's time, Pliny wrote to rather over a hundred. Most of these hundred or so persons were senatorials or equestrians. They had money, they had country houses and their investment (like Pliny's own) was often in land. So Pliny sings the praises of Caninius Rufus' house at Comum (I. 3*), while from the very next letter we learn that Pliny's mother-in-law (Pompeia Celerina, the mother of his second wife) had houses at Ocriculum, Narnia, Carsulae and Perusia (I. 4*). Elsewhere, Suetonius is on the look-out for a small estate in the country (I. 24*). Country life seems to have appealed to these people, at least in small doses. From Pliny, but from nowhere else, we learn the perhaps surprising fact that Tacitus was in the habit of going on boar-hunts — and not, as Pliny did, to catch up on his reading and writing (I. 6; IX. 10*).

But, as well as the country, Rome is never very far away in Pliny's letters. In I. 9, Pliny speaks of the routine of commitments in the capital: a coming-of-age ceremony; a betrothal; a wedding; witnessing a will; supporting someone in court or acting as an assessor. Pliny is talking of himself; but such a programme was presumably common among his friends and acquaintances. Then there was political life. So (to take one example) Pliny finds himself canvassing for Sextus Erucius, as he stands for tribune (II. 9). And there was the life of the courts (IV. 16, for instance). The days were crowded in the city, even if they were perhaps not quite as crowded as Pliny made out to Caninius Rufus (II. 8).

Whether the letters are set in the city or the country, the overall picture is of a very stable, self-assured society. Pliny and the people with whom he mixed were privileged. They had been born to positions of authority and leadership, whether that leading place was in the running of the empire or in an Italian town or as the owner of great estates. Pliny for one accepted the value of the distinctions of class and rank. As he told Calestrius Tiro on one occasion: 'Once these are mixed up together and thrown into confusion, there is nothing more unequal than the ensuing equality' (IX. 5*). But nowhere in Pliny's letters is there a fear that the structure of society is about to collapse. He even jokes with his mother-in-law about the casualness of his own slaves (I. 4*); hardly the attitude of a worried man; and, though he is shocked by the news that Larcius Macedo has been murdered by his slaves, there is the feeling that this murder took place not because the structure of society was at risk but because Macedo had been a proud and cruel master (III. 14). In another letter to Calestrius Tiro, Pliny speaks of the

country being in a flourishing state ('florente republica', I. 12*). And
this feeling of confidence pervades the Letters.

It is good that Pliny gives so clear a picture of the time. Too often
the privileged end of Roman society reveals itself only from a distance.
But Pliny meets leading Romans at close quarters in his letters and
shows them in their day-to-day work and leisure. Inevitably, since he is
not concerned with a complete cross-section of society, his is an incom-
plete picture. It may also be idealised and with too great an emphasis
on calm. In the same way a Gainsborough portrait of an English aristo-
crat gives a deceptively serene impression and represents only one part
of eighteenth-century life. But, for all that, Pliny brings himself and his
peers out of the pages of the history books and down from the bare
inscriptional records and makes them real people.

(ii) Book X

Book X is completely different from the other nine Books. All but the
first fourteen letters of this final Book are an official exchange of
letters between Pliny, as Trajan's special envoy in Bithynia, and the
Emperor. They were published only after Pliny's death; and Pliny wrote
them, not as self-consciously literary pieces, each elegant phrase honed
to perfection, but because he needed to write to the Emperor on some
problem or other in his governorship of Bithynia. The replies (rescripts),
written by Trajan or a member of his secretariat, were equally businesslike.[9]

The letters in Book X are in fact the raw material of history. A historian, however impartial he may try to be, will necessarily be subjec-
tive to some degree or other from the moment he chooses what to
write about; and his writings will be, however slightly, imprinted with
his personality. However, what Pliny and Trajan wrote to each other
was history in the making: in the wake of these letters action was taken
that affected the Bithynians; and the letters, the only surviving set of
government correspondence of this kind, give a unique and objective
glimpse of the day-to-day running of a great empire.

Bithynia

The province of Bithynia and Pontus stretched eastwards from the
Bosphorus along the south coast of the Black Sea. The only part of the
province on the European side of the Bosphorus was the city of Byzantium. As its double name implies, the province was once two territorial
areas. Bithynia had been a kingdom for a long time before it became

---

9 For a discussion of how much hand Trajan had in the rescripts to Pliny,
   see particularly S-W, 536-46, discussed briefly in *CLC Handbook* for
   Unit IV, p. 64.

part of the Roman empire in 74 BC, when the last king of Bithynia, Nicomedes IV, bequeathed it to Rome. At this time Rome was experiencing trouble with Mithridates VI in the East. Pompey the Great was appointed to the command against him in 66 BC. In 63 BC Mithridates committed suicide; and Pompey, as part of his general settlement of the eastern Roman empire, incorporated the western part of Pontus (which lay to the east of Bithynia) into a new province of Bithynia and Pontus. In 27 BC responsibility for administering the provinces of the Roman empire was divided between Augustus and the senate. In this division Bithynia and Pontus became one of the senate's provinces; and in common with the other senatorial (public) provinces its governor was chosen by lot (*sorte*), was called *proconsul* and governed normally for one year.[10]

In Bithynia life had long revolved round the Greek cities and this continued under the Romans. Time and again Pliny's letters to Trajan concern individual cities, the capital city of Nicomedia, Nicaea, Prusa, Byzantium and so on. The countryside was part mountainous, part a land of fertile plains. In Bithynia's early days as a province the poet Catullus had served there on the staff of the governor. He afterwards wrote of his joy at being home again at the family home at Sirmio in the north below the Alps: 'scarcely believing that I have left Bithynia and its plains and see you safe and sound' (XXXI, in *CLC* Unit IV).

Over the next one hundred and seventy years or so, till Pliny's appointment to the province, Bithynia and Pontus made little impact on events.

Pliny's appointment to Bithynia

The Milan inscription recording Pliny's career styles him '*legatus pro praetore* of the province of Pontus and Bithynia sent with consular power into that province as a result of a *senatus consultum* by the Emperor' (Appendix no. 1). The words of the inscription need some explanation. Pliny had not been appointed in the normal way for Bithynia, by lot: he owed his appointment to the Emperor. In addition, he was not called *proconsul* (the normal title of the governor of Bithynia), but *legatus pro praetore*. This further emphasised his direct responsibility to the Emperor, because in the provinces allocated to the Emperor, each governor was traditionally called *legatus Augusti pro praetore* (see S-W, 82-3).

It is clear then that Pliny's appointment to Bithynia was a special one. This was not the first time an emperor had stepped in to appoint a governor to a public province. It had, for instance, happened in Cyprus

10 See Notes on VII. 16 for the administration of the provinces during the principate.

in the principate of Augustus;[11] but it was sufficiently unusual to indicate that Bithynia and Pontus had special problems.

The exchange of letters between Pliny and Trajan, as well as Pliny's own career, suggest what these problems were. Pliny was a man with no military experience, except as a young man in Syria where he was put to auditing accounts. He had never served abroad after that early military appointment. So the problems of Bithynia were not those that called for a soldier or a man with previous experience of provincial government. Pliny, however, already knew something about the province, having defended two ex-governors of Bithynia during the previous ten years, and this may have influenced Trajan in appointing him. But his special skill was finance (see section II: Pliny's career); and the finances of the cities of the province were in a mess. Time and again cities had started public works but had failed to complete them. The citizens of Nicomedia had spent 3,318,000 sesterces and then 200,000 sesterces on aqueducts that had been abandoned (X. 37). Pliny found the theatre at Nicaea only half built, though it had already cost more than 10,000,000 sesterces, so he had been told (X. 39*). And so on. Trajan clearly did not want a cut-back in expenditure on public works (note the closing of the open sewer at Amastris: X. 98/9, and the Emperor's final consent to Pliny's scheme for a canal: X. 41/2 and 61/2), but he did not want public money misused. There was also at least the fear of public disturbance in the province. Following the Emperor's instructions (*mandata*) Pliny issued an edict forbidding political societies (*hetaeriae*: X. 33/4* in *CLC* Unit IV 'Bithynia' and see *Handbook*, pp. 77ff.). Probably too, with such fears of public disturbance at the back of his mind, Pliny dealt more harshly than he needed with the Christians, while Trajan seems more tolerant (X. 96/7). There were also apparently intrigues among the governing class. But the misuse of public money seems to have been the main reason for the choice of Pliny.[12]

How successful was Pliny in Bithynia?
Finally, there is the question of Pliny's success as governor of Bithynia. It has been fashionable to laugh at Pliny's letters to the Emperor from Bithynia and to accuse him of indecision. The final letter in this selection (X. 117) seems to back this up; it finds Trajan telling Pliny: 'I made you my choice so that ... you could make your own decisions about what is necessary for their [the Bithynians'] peace and security.' However, a study of the complete exchange of letters between Pliny and Trajan may persuade the reader to a different conclusion. First, though Pliny was the Emperor's special appointment and had, as he

11 P. Paquius Scaeva was appointed proconsul of Cyprus 'iterum extra sortem auctoritate Aug. Caesaris et s.c. misso' (*ILS* 915).
12 See S-W, 527 for the problems in Bithynia.

said, Trajan's authorisation 'to refer to you about matters in which I am in doubt' (X. 31) — an injunction which was perhaps included in his *mandata* — he only wrote back to Rome about once a fortnight. This by modern standards seems almost casual. As to the letters themselves, public works loom large in them (as has already been implied). This shows that Pliny was not prepared to be a negative governor. Where new public works were required, he got things going, even on one occasion writing to ask Trajan to part with a piece of imperial property to the city of Prusa so that a set of public baths could be built on the site (X. 70*). This concern for the welfare of the people of the province fills Pliny's letters. Having banned *hetaeriae*, he knew that he could not legally form a much-needed fire-brigade in Nicomedia; but, rather than let matters rest, he wrote to try to persuade the Emperor to make an exception in this one case (X. 33* in *CLC* Unit IV, 'Bithynia'). He also tried to get Trajan to deal leniently with certain public slaves, though as a lawyer he knew that the law called for a harsh line (X. 31). Though Juliopolis was a town in the back of beyond, he listened to its citizens when they complained about being put upon by Romans travelling through, and wrote to ask Trajan to send soldiers to police the town (X. 77). Of course, there were the occasional times when it seems to the reader that Pliny vacillates a little. But he had a mind trained to spot legal niceties and this probably made him see all sides of a question a little too clearly (X. 116). The overall impression, however, is of a positive, benevolent governorship (except possibly in the case of the Christians: X. 96). Trajan does not always acquiesce in Pliny's requests; but the splendid thing is that both Emperor and governor show an unmistakable concern for the interests of the Bithynians. Trajan, one suspects, rated Pliny a success in Bithynia.

IV Pliny's 'Letters' as literature

The title of 'literature' seems to be bestowed on writings which survive over generations because their content is interesting and their style of composition is a pleasure to read.

Their content alone would make Pliny's letters an indispensible reference work for historians, for they contain a wealth of material inaccessible elsewhere. They give a vivid picture of life in governmental circles under the Emperor Trajan and his immediate predecessors and tell us a great deal of how the imperial system worked at the level of actual human relationships (see II. 9, IV. 22, VII. 29, and the letters in Book X). There is not much discussion on serious questions of politics, but there is plenty of graceful description and anecdote, spiced with a friendly humour (e.g. I. 13, III. 14, VII. 24). There are few insights into the darker side of social life or the lot of underprivileged people, but we do learn how the humane culture of Greece had profoundly affected the outlook of some of the Romans in public life; and the letters to the Emperor (Book X) give us a glimpse of how a civilised Roman behaved, and on what principles he acted, when put into a position of almost absolute authority.

However, the style of the Letters has ensured them, over the centuries, a wider value than simply as material for the historian, and given them a real claim to the title of literature. The writing does not have the depth or variety of Cicero, but it is lucid, easy to read and relaxed, in spite of the frequent mannerisms of the rhetorician. It is the Latin of a man thoroughly versed in the extravagant rhetorical methods of his time who has decided to cultivate, for this particular *genre* of writing, a more intimate and simple style. In VII. 9* he says to a friend, encouraging him to write letters, 'pressus sermo purusque ex epistulis petitur'. Letter-writing, he explains, is a salutary change from the hard, aggressive style demanded by public oratory.

It is admittedly debatable how much Pliny himself fulfils the promise of 'pressus sermo purusque' in his own letters. Sherwin-White (p. 5) is decided on the matter: 'Passages may be taken, especially from the opening of letters, which at first sight promise simplicity, but artifice is soon apparent in the construction, if not in the vocabulary . . . the cunning of the artist's hand appears in the order and arrangement.' But there is another point of view here: it is arguable that what appears to be artifice is really the style which had become natural to Pliny from his rhetorical training and then his long years of composing attorney's orations, official addresses and speeches in the senate; and that any cunning which Pliny employed in composing the letters consisted rather of trimming and simplifying the extravagant and verbose periods which by this time came automatically to him. We must not underestimate the thorough immersion of a man like Pliny in 'the great, roaring machine of Latin rhetoric' (C. S. Lewis, *Selected literary essays*, p. 127).

Whatever the truth of how the style was achieved, it is a pleasing style. But how much does it owe to earlier writers? Literary echoes do recur, and, as Sherwin-White points out (pp. 16-18), various people have been cited as influencing Pliny — including Cicero, Statius, Martial and Horace. Pliny has even been accused of adopting whole themes from earlier writers and dealing with them in a similar style: for example, VIII. 17, on the flooding of the Tiber, has been compared to Horace's *Odes* I. 2. However, there is a limit to how seriously we can take this kind of thing as evidence of imitation: Pliny is after all seeing the same things as Horace saw, and it would be difficult for the description not to contain similarities. Elsewhere too we can find phrases reminiscent of earlier writers, but this is what we expect in the style of any man who has been educated in his own country's literature. A more curious phenomenon is the letter VI. 16, where Pliny, writing about the death of his uncle under Vesuvius, is presenting the story to Tacitus for possible

use in his history: the odd thing is that Pliny adopts for much of the
letter (especially line 14 onwards) a clipped, elliptical style highly
reminiscent of Tacitus himself — almost as if he is not only giving
Tacitus the story, but also putting it into the right style for immediate
use.

So it is difficult to pin down any earlier writer to whom Pliny is in-
debted for his style; and it is even more difficult to find a predecessor
in his particular *genre* of letter-writing. Cicero of course springs to mind,
but he was not writing for publication. At most, some of his letters were
intended to be circulated as political documents, and these show great
formality and elaboration; but the majority were written simply to give
current news to an intimate friend or to pour out the writer's feelings,
and their style is brief, idiomatic and full of slang and allusion. There
are not many points of contact between Cicero and Pliny. More promising is a comparison with Horace's *Epistles,* or the occasional pieces
to friends written by Statius and Martial — all of these in verse, but far
nearer to the kind of thing which Pliny is attempting. It is interesting
to note that Martial was one of Pliny's own protégés (cf. *CLC* Unit IV
'domi' and *Handbook,* p. 151).

Whether the letters are a collection of genuine correspondence with
other people — as they purport to be — is a matter more for speculation
than for debate. The extreme view, that the letters are largely fictitious,
is absurd: to invent the kind of detail contained in them, and to weld
such invention into such a consistent whole and make it so realistic
would have required a *tour de force* of imagination and labour which in
any case was quite unnecessary: why should Pliny have done such a
thing when he must have had a wealth of private correspondence, existing in copies or in his memory, on which to base his publication? Clearly
we must assume that not only the letters to Trajan in Book X, but the
other letters too, have their basis in authentic correspondence: but
then arises the question of how far that correspondence has been
edited or rewritten.

Here we really are in the realm of speculation. We do not know in
what style Pliny composed the originals: was he, even then, writing to
friends in a deliberate style of 'pressus sermo purusque', for their pleasure, or with a view to publication, or from his own desire to practise
that style — or with all these motives? Was he in the habit of writing to
friends the interesting essays on current events such as III. 14, IV. 2 or
IV. 22? Or are these merely parts of longer letters, extracted and
fashioned into polished pieces? In short, how much editing has occurred? Such questions can be asked about even the letters which are
more concerned with business and personal affairs, such as I. 19, II. 9,
III. 6 and IX. 39, where the lack of more trivial, mundane details
suggests careful pruning. Likewise the love-letters to Calpurnia look
like extracts from longer letters which must have contained all kinds of

news and gossip. Perhaps we see the reverse kind of process in the editing of I. 19 and IV. 1, where the recipients of the letters are being given information which they could not fail to know already: Pliny has presumably wished to include these letters in his selection, but in order to make them intelligible to the reader has had to add a certain amount of detail not in the originals.

It would be dangerous to go beyond these fairly obvious examples of editing, nor would we be justified in assuming that the letters which give the strongest impression of formality are the ones which received the most alteration before they were published. We can only speculate on how Pliny arrived at the collection as we now have it. All we can say is that the letters have clearly been chosen for their variety of interest, but they possess a unity which is achieved by consistency of style and form.

How did Pliny himself esteem his letters? Did he regard them as his most important attempt at literature? This is a fascinating question. He does make it clear that his ambition is to write literature: he says that he seeks inspiration by the seashore at Laurentum, which is 'verum secretumque μουσεῖον' (I. 9); and he hopes that his work is going to survive after he is dead: 'occursant animo mea mortalitas mea scripta . . . enitamur ut mors quam paucissima quae abolere possit inveniat' (V. 5). What we do not know is whether he is here referring to his speeches, some of which he certainly published, or to his letters. It is possible that as a public man he had at first regarded his speeches as the works most likely to survive, but later, on the advice of his friends or on his own judgement, he realised that the letters might be more esteemed by posterity. If so, he was correct, for his only surviving speech, the elaborate *Panegyric* on Trajan, has long been relegated to the reference-shelves, while the letters have been read with delight ever since their publication.

V Pliny the man

Pliny's *Letters* reveals a wealth of information about the society and history of his time. But it also reveals a great deal about Pliny himself. Too few people from past ages come down to us as anything like complete personalities. This is not a deficiency unique to the ancient world of Greece and Rome. In the English history of the fifteenth century, for instance, the kings and great nobles speak to us chiefly through their public lives, their politics and their battles, and it is difficult to hear them as real people or to see them at home with their families. The Norfolk family of Paston or the wool-merchant Thomas Betson come across as clearer, fuller personalities than, say, Edward IV or his father, the Duke of York. The reason for this is relevant to Pliny: the letters of Thomas Betson and of the Pastons have survived, whereas those of the great nobility of the fifteenth century have, in the main, not.

Books I-IX, even though they were revised by Pliny himself before publication, give a remarkably complete picture of their author. 'The letters,' Mrs Radice has remarked, 'are more than a source book; they also paint the fullest self-portrait which has survived of any Roman, with the possible exceptions of Cicero and Horace' (*The Letters of the Younger Pliny*, Penguin, p. 27). At every turn of the page Pliny reveals some detail of his daily life or displays some attitude or prejudice. We come to know him very well indeed. We can almost hear his voice.

One obvious aspect of Pliny is his intense interest in literature (see above, pp. 9-10). Two letters not so far mentioned add something to a consideration of Pliny and writing: I. 13, where he deplores (perhaps a little pompously) the poor standard of behaviour at *recitationes* (he accuses people of 'desidia vel superbia' for not attending *recitationes*, even though they must often have been boring occasions to those less dedicated to literature than Pliny), and VII. 21, where, suffering from eye-trouble, Pliny tells Tertullus: 'I am neither writing nor reading — no easy sacrifice.' These few words to Tertullus sum up the consuming appeal of literature to Pliny.

Reference has also been made to Pliny as a patron (pp. 10-11). Both communities and individuals benefited from his patronage. In addition to letters already mentioned, he writes to Mustius about his proposed rebuilding of the Temple of Ceres on his estates (IX. 39), and to Annius Severus about the gift of a Corinthian statue to his native district (III. 6). This second letter tells us not merely about the workings of patronage: it shows, too, Pliny's affection and concern for the area where he was brought up (Comum and its surrounding countryside) and for the people there. There is plenty of other evidence for this sense of *patria*. In IV. 13* he proposes helping to provide Comum with a schoolmaster. He built and endowed a public library there (I. 8*); he provided financial assistance (*alimenta*) for children in need (VII. 18*); and having spent perhaps about 2,000,000 sesterces on benefactions to Comum during his life-time, he gave the town more than 2,000,000 sesterces by the terms of his will (Appendix no. 1).

Individuals, both at Comum and elsewhere, received Pliny's help. Firmus, to whom he proposed giving 300,000 sesterces, was a town-councillor of Comum (I. 19). Then there was the family of Corellius Rufus. Rufus had helped Pliny considerably in his early days: so Pliny in his turn acted for the family where he could, obtaining a tutor for Corellius' grandson (III. 3*), representing Corellius' daughter in court (IV. 17*) and selling land cheap to his benefactor's sister (VII. 11* and 14*). In the case of Suetonius, he did his best to obtain a property for him at an advantageous price (I. 24*); and later he wrote to the

Emperor for privileges for him (X. 94/5). Of course patronage, *amicitia* and *officium* were social, almost moral duties to a Roman. But Pliny's letters leave us with the impression that he was generous beyond what was required of him.

The letters are permeated by his warm-heartedness. What he writes is shot through with a humane and compassionate insight into other people's lives and feelings. His treatment of freedmen and slaves was exemplary. He was kindness and generosity itself to his freedman Zosimus, who suffered from persistent ill-health (V. 19). He had his freedmen sharing the same table with him at meals (II. 6), and in his view most of the accommodation reserved for his slaves and freedmen in his villa at Laurentum was presentable enough to be used for guests (II. 17*). He asked Calestrius Tiro, who was travelling out to be governor of Baetica, to turn aside from his journey to his province to manumit formally some of his wife's grandfather's slaves (VII. 16). After Larcius Macedo's death, he felt both for Macedo and for the slaves who had murdered him (III. 14). But particularly significant is his remark (VIII. 16) that men who see the deaths of their slaves only in terms of monetary loss 'are not human beings'. Pliny's attitude to slaves bears comparison with that of Seneca a generation or so earlier. In one letter Seneca wrote that he had slaves at table with him; and at the start of this same letter (an eloquent manifesto of Seneca's humane view of slaves) he wrote: '"They are slaves", people say. No, they're human. "They are slaves". No, they're comrades. "They are slaves". No, they're humble friends' (*Letters*, XLVII). Seneca's Stoic philosophy had influenced his treatment of slaves; and such treatment by Seneca and his fellow-Stoics may have influenced Pliny, though it should be said that Seneca and Pliny were still probably exceptional in the way they looked on slaves.

Pliny's warm-heartedness extended far beyond his slaves and freedmen. He understood that young people had feelings. He told Terentius Iunior to be more tolerant towards his son ('Remember . . . that you have been a boy too, and use your rights as a father without forgetting that you are only human and so is your son', IX. 12). He deplored Regulus' attitude to his son, both while the child was alive and also after his death (IV. 2) — but then, he had a personal dislike of Regulus.

But it is in his attitude to his own family that Pliny emerges at his most attractive. He writes with an obvious warmth to his third wife Calpurnia's grandfather, Calpurnius Fabatus (IV. 1, for example). Meanwhile, his letters to Calpurnia are real love-letters, particularly VII. 5: 'It is unbelievable how much I miss you', he starts, telling her how he visits her empty room at the times he normally visits her there, turning aside sick and sorrowful and like a lover locked out. Other letters tell of Calpurnia's miscarriage (VIII. 10* and 11*) and of her inability to have any more children. The last of Pliny's letters, in Book

X, perhaps the last he actually wrote to Trajan (X. 120* in *CLC* Unit IV 'Bithynia'), reveals that Calpurnia has gone back to Italy to be with her aunt after her grandfather's death. This was a final parting from Pliny, who is believed to have died shortly afterwards.

The preceding pages have tended to concentrate on Pliny's many good qualities: his warm-heartedness, his humanity, his capacity for affection, his liberal attitudes in a time that was in many ways illiberal. But he certainly had his faults. Several of these are easy to see. Some of his contemporaries criticised him for exaggerated praise of his friends (VII. 28*). And it is true, he had a great capacity to flatter. Tacitus often benefited from this. On one occasion, for instance, Pliny told Tacitus: 'It was not as one master to another, nor as you say, as one pupil to another, but as a master to his pupil (for you are master, I am pupil . . .) that you sent me your book' (VIII. 7*). But on this occasion at least it looks as if Tacitus had himself been going in for a little flattery. Pliny is also open to the charge of priggishness — read, for instance, the letter where he deplores the way his host grades his guests (II. 6). He is, for all his liberal attitudes, a snob: he reckons that the crowds (*vulgus*) at the chariot-races are more worthless than the coloured tunics they back (IX. 6). Snobbishness also makes him inveigh against Pallas, the long-dead influential freedman of Claudius; but in the end his sense of proportion makes him laugh at his own indignation (VII. 29). He sometimes gives the impression that he believes he knows all the answers. Though he had not yet been a provincial governor, that did not stop him giving words of advice to Maximus, Trajan's special envoy to Achaea (VIII. 24*), or again to Calestrius Tiro both before he went out to be governor of Baetica (VI. 22*) and later on (IX. 5*) when he wrote with congratulations and further advice. Were his contemporaries infuriated by this kind of behaviour? Perhaps not, because *amicitia* included an obligation to give advice. Akin to this avuncular attitude is Pliny's distinct lack of modesty: 'It has often happened to me when speaking in the Centumviral Court that my hearers have . . . suddenly jumped to their feet with one accord to congratulate me as if driven by some compelling force' (IX. 23*). Or again, to Tacitus: 'Whether posterity will give us a thought, I do not know, but surely we deserve one' (IX. 14*). This concern with fame was almost an obsession with Pliny. 'Pliny,' says Syme, 'was frank and exorbitant in his demands upon fame' (*Tacitus*, I. 113). In Syme's view he wrote his *Letters* as his own *Res Gestae* and, by publishing them in his own lifetime, he 'stole a march on the funeral oration' (*Tacitus*, I. 98). No doubt his other works too, his speeches and poetry which he prepared for publication, were intended to win him fame. Such immodesty may ruffle a reader's puritan feathers. But a preoccupation with fame was commonplace among ancient Romans. Fame was a passport to immortality. Tacitus saw Agricola surviving through his biography (*Agricola*, 46); and he can therefore be taken to have hoped that his writings would bring immortality to himself also.

Among other criticisms of Pliny there is sometimes a certain impatience with his middle-aged attitudes. Some readers wish, for instance, that he would go off on a sexual orgy or two and then come back and describe them. There is a danger here that we may be expecting the wrong things of Pliny. Pliny must be taken for what he is and judged fairly: he must not be blamed for failing to provide for his readers exciting vicarious experiences or censured because he is at times too accurate a mirror of his readers' lives.

But perhaps more damaging criticisms of Pliny are possible. A sense of elegance, of gentleness and amiability informs Pliny's letters. But this may only be a veneer, with Pliny consequently being open to the charge of being superficial as a writer. To pinpoint superficialities, the letters seem heedless of the dark years of Domitian's reign, when Pliny rose successfully through the *cursus honorum* towards the consulship. Again, there are some charming pen-portraits of individuals, but the letters do not contain much in the way of character-study. Nor are there any penetrating moral judgements. Even an admirer may sometimes wonder, does all this indicate a shallow intellect in Pliny?

In seeking to examine the charge against Pliny of superficiality, we might do worse than compare him with Tacitus. After all, the contrast between their writings is striking. The comparison is between two men who were contemporaries and friends, who were both in public life and were both writers. Tacitus was about six years Pliny's senior, being born in AD 55. Like Pliny he pursued a successful political career under Domitian and he became *consul suffectus* in AD 97, three years before Pliny's own suffect consulship of AD 100. As a leading lawyer he collaborated with Pliny in the trial of Marius Priscus. (See I. 6, under 'The letter', for further details of Tacitus and Pliny.) But, to take up particular points already mentioned, where Pliny seems superficial, there is in Tacitus no lack of awareness of Domitian's reign; there is abundant evidence of probing character-study; there is no shortage of moral judgement; and, at the end if not long before, a reader becomes conscious in Tacitus of a man with a formidable intellect.

It is however essential to the argument to realise that Pliny and Tacitus were not writing the same type of literature. Tacitus was a historian, whereas Pliny was writing *belles-lettres*. As a historian, Tacitus was by definition writing about the past and could therefore be expected to be aware of, and even to dwell on, Domitian. (The latter part of his *Histories* covered Domitian's principate; and it was under Domitian that Agricola spent several years of his governorship of Britain before his recall.) Pliny's letters deal with the present, largely with Trajan's reign. Pliny and Tacitus were also writing about rather different (even if sometimes overlapping) aspects of Roman life. Pliny was concerned with his own social milieu, writers, landed gentry, men of the law, politicians in the daily minutiae of their lives. But Tacitus chose to write about power at the very centre and life at the court of the emperors. These two different areas of life inevitably called for different treatment. Character-study and moral judgement would come more easily to Tacitus in view of what he was describing, while the pursuits of a wealthy, privileged class would encourage a less intense, less censorious form of writing. A historian is expected to go probing into human behaviour and to analyse the characters of the leading people of the past. If he romanticises that past, he may be called popular (particularly nowadays, if his books sell). Perhaps a writer of *belles-lettres* however has a duty to romanticise a little. In short, the nature of the literature each was writing could contribute to the impression that Tacitus was intellectually the greater of the two men. Yet a nagging thought remains: there have been writers of literature akin to Pliny's Letters who go below the surface. Samuel Pepys has an acuter observation of human behaviour, Seneca a more philosophical turn of mind and Montaigne more developed gifts as a moralist. The charge against Pliny of superficiality cannot be made to stick on all counts; but there may be more than a grain of truth in it.

In the meantime the comparison between Pliny and Tacitus raises the question of Pliny's temperament. Tacitus' writings reveal a pessimistic view of the world, Pliny's Letters an optimistic view. The pessimism and the optimism are each in their way a compliment to the new, happy times of Trajan's principate. But they may too be sign-posts to the characters of the two men, who pursued successful political careers under Domitian but have left such different impressions of that time. It made a searing impression on Tacitus, who remembered with horror the fifteen years when Domitian was emperor, 'no small part of a man's life' (*Agricola*, 3); and it seems to have coloured his view of all emperors: 'Tacitus was obsessed by the real or imaginary Domitians of past history' observes Michael Grant (*Tacitus, The Annals of Imperial Rome*, Penguin, 18). Pliny also refers to the dangers of Domitian's reign: 'I stood amidst the flames of thunderbolts dropping all around me, and there were certain clear indications to make me suppose a like end was awaiting me' (III. 11*). But his lack of obsession with the dangers makes one wonder if he is merely paying lip-service to the general senatorial hostility to Domitian. It is of course dangerous to draw conclusions about a person's character from his writings; but the writings must be allowed to provide a few clues. Perhaps then Tacitus was a tortured, sensitive soul, marked for life by what he saw as his own political inadequacies in Domitian's reign, a man relentlessly seeking in others his own shortcomings. Pliny on the other hand seems to have been far more equable (his critics would say insensitive); and his success may have made an already optimistic nature even more optimistic, the result being that amiable view of the world that is one of the most attractive features of his Letters.

And it is to the Letters that we always have to return in the end, for that is where Pliny is to be found. There is however a caveat. What we read there is, with the important exception of Book X, what Pliny wants us to read. The picture both of Roman life and of Pliny himself is a selective picture, with Pliny making the selection. The picture of the contemporary scene is probably accurate enough: after all, Pliny would have lost credibility with his readers if he had given a distorted view — he never pretended to be a satirist. But how much is missed out of the full picture of Pliny? There is precious little in what he says to suggest that he could ever have survived in the harsh world of Roman politics as depicted by Tacitus. Yet he not only survived but also prospered, this benign and often complacent lawyer-cum-country-gentleman who was a model patron, who took endless delight in his books and who retired whenever possible to his country-house in relief to be away from the rat-race of Rome. The Milan inscription (Appendix no. 1) corroborates the impression of Pliny's generosity gained from his Letters; many of the letters of Book X confirm Pliny's concern for the well-being of his fellow human-beings. But his letter to Trajan about the Christians (X. 96) shows a toughness towards the Christians that may indicate a toughness in Pliny in other dealings too, say in his political and legal career. Pliny must have been tough to have made such a success of his life. But the leisurely and scholarly Pliny we see so often in the Letters will have pulled very hard at times and often urged on him a life of quiet. A letter to Valerius Paulinus (IX. 3*) may provide the clue to this dichotomy in Pliny; in it he writes: 'Were my own eyes not fixed on the reward of immortality, I could be happy in an easy life of complete retirement.' On the evidence of his career and of his *Letters,* this one sentence may perhaps better than any other sum up Pliny as a person.

SELECT BIBLIOGRAPHY

This is not intended as an exhaustive bibliography but as a list of books and articles which we have found particularly useful when writing this *Handbook* and which we believe others can consult with profit.

1. PRIMARY SOURCES

a. Literary
Juvenal, *Satires*
Martial, *Epigrams*
Pliny, *Panegyric*
Seneca, *Moral Letters*
Suetonius, *Lives of the Twelve Caesars*
Tacitus, *Annals, Histories, Agricola, Dialogus*

Most of these works have been translated in the Penguin Classics (for availability, consult current list of Penguins in print). The introductions to these translations are often of considerable value, in particular Michael Grant's introduction to his *The Annals of Imperial Rome*, rev. edn (Penguin, 1971).

b. Inscriptions
Corpus Inscriptionum Latinarum (Berlin, 1863-)   (*CIL*)
Inscriptiones Latinae Selectae, ed. H. Dessau (Berlin, 1892-1916)
    (*ILS*)
Documents illustrating the reigns of Augustus and Tiberius, collected
    by Victor Ehrenberg and A. H. M. Jones. 2nd edn (Oxford U.P.,
    1955)   (EJ)
Documents illustrating the principates of Gaius, Claudius and Nero,
    ed. E. Mary Smallwood (Cambridge U.P., 1967)   (Smallwood (1))
*Select documents of the principates of the Flavian Emperors including
    the year of the revolution, A.D. 68-96*, ed. M. McCrum and A. G.
    Woodhead (Cambridge U.P., 1961)   (MW)
Documents illustrating the principates of Nerva, Trajan and Hadrian,
    ed. E. Mary Smallwood (Cambridge U.P., 1966)   (Smallwood (2))
The Roman Inscriptions of Britain, ed. R. G. Collingwood and R. P.
    Wright, Vol. I (Oxford U.P., 1965)   (*RIB*)

We have made considerable use of inscriptions in view of their central
importance as source material. There is no simple introduction to the
study of inscriptions; but Graham Webster's *A short guide to the
Roman inscriptions and sculptured stones in the Grosvenor Museum,
Chester* (rev. edn 1970, available from the Grosvenor Museum) provides
an interesting and helpful starting-point. Of immediate help to the
understanding of inscriptions is a visit to a museum containing a good
epigraphical collection, such as Bath or Newcastle.

2. MODERN WORKS

a. Pliny
Hardy, E. G. (ed.) *Pliny's correspondence with Trajan* (Macmillan, 1889)
Sherwin-White, A. N. *The Letters of Pliny: a historical and social
   commentary* (Oxford U.P., 1966)   (S-W)
Syme, Sir R. *Tacitus* (two volumes) (Oxford U.P., 1958)   (Syme,
   Tacitus)
Radice, B. (tr.) *The Letters of the Younger Pliny* (Penguin, 1963, re-
   printed with bibliography 1969)

Hardy's commentary is important for an understanding of the letters
of Book X. Sherwin-White is indispensable, as almost every page of this
Handbook shows. Syme's *Tacitus* has illuminating chapters (VII and
VIII, vol. I, 75-99) on Pliny's career and writings (note also Appendices
19, 20, 21, 25, 27, 28). The book is also a mine of information about
Pliny's contemporaries and the Rome of his time. Mrs Radice's introduction to her Penguin translation is a useful summary, while the
translation itself is careful and pleasant to read.

b. General
Balsdon, J. P. V. D. *Life and leisure in ancient Rome* (Bodley Head,
   1969)
Balsdon, J. P. V. D. (ed.) *Roman civilisation* (Penguin, 1969)
The Cambridge Ancient History Vols. X and XI (Cambridge U.P., 1934
   and 1936)   (*CAH*)
Carcopino, J. *Daily life in ancient Rome* (Penguin, 1956 (Carcopino)
Crook, J. A. *Law and life of Rome* (Thames and Hudson, 1967)
Grant, M. *The world of Rome*, rev. edn (Cardinal, 1974)
Oxford Classical Dictionary 2nd edn (Oxford U.P., 1970)   (*OCD*)
Paoli, U. E. *Rome, its people, life and customs* (Longmans, 1963)
   (Paoli)

Two chapters of *Roman civilisation* are of particular relevance to this
Handbook: chapters 4 ('The last crisis: the Roman Empire to its decline'
by A. H. M. Jones) and 5 ('Roman imperialism' by A. N. Sherwin-White).

c. Individual topics
Dudley, D. R. *Urbs Roma* (Phaidon, 1967)
Jones, A. H. M. *The cities of the eastern Roman provinces* 2nd edn (Oxford U.P., 1971)
Jones, A. H. M. *Studies in Roman government and law* (Blackwell, 1960)
Lepper, F. A. *Trajan's Parthian War* (Oxford U.P., 1948)
Millar, F. 'Emperors at work' *JRS* LVII (1967) 9-19
Millar, F. 'The Emperor, the senate and the provinces' *JRS* LVI (1966) 156-66
Millar, F. *The Roman Empire and its neighbours* (Weidenfeld and Nicolson, 1967)
Sherwin-White, A. N. *Roman society and Roman law in the New Testament* (Oxford U.P., 1963)
Wheeler, Sir M. *Roman art and architecture* (Thames and Hudson, 1964)
Wirszubski, Ch. *Libertas as a political idea at Rome during the late Republic and early Principate* (Cambridge U.P., 1950)

Dudley's *Urbs Roma* is 'a source book of classical texts on the city and its monuments selected and translated with a commentary'. It also contains 105 photographs of Rome and its classical remains.

Jones' *Studies in Roman government and law* contains one particularly valuable chapter: chapter I, 'The Imperium of Augustus' (first printed in *JRS*, XLI, 1951), outlines Augustus' *de iure* powers and is therefore of relevance to an understanding of the powers of future emperors. Millar's 'The Emperor, the senate and the provinces', which is not just confined to Augustus, adds to this discussion of the powers of the emperors and sets Pliny's appointment to Bithynia in its historical context.

Millar's *The Roman Empire and its neighbours* covers the years AD 14 (the death of Augustus) to AD 284 (the accession of Diocletian). The Younger Pliny appears frequently in this straightforward, helpful book.

Sherwin-White's *Roman society and Roman law in the New Testament* is his Sarum Lectures of 1960-1 in book form. The book is too little known: in it Sherwin-White dovetails the early Principate in with the early Christian church and in the course of his lectures does much to clarify all sorts of topics such as *imperium* (1ff.).

Wirszubski's *Libertas* is the definitive discussion of what Romans meant in political terms by the word *libertas*.

d. CLC Teacher's Handbooks, especially those for Unit III (Cambridge U.P., 1973) and for Unit IV (Cambridge U.P., 1976)

e. Cambridge Latin Texts
Jones, P. V. *Selections from Tacitus Histories I-III: The year of the four emperors. Handbook.* (Cambridge U.P., 1975)

COMMENTARY ON THE LETTERS

I. 6

The letter

Pliny writes to his friend Tacitus about a day's hunting, an occasion which he has also used for some quiet study.

This is one of eleven letters which Pliny wrote to the historian Tacitus. Others include the two famous letters giving Pliny's eye-witness account of the eruption of Vesuvius, the death of his uncle and the rescue of his mother (VI. 16; VI. 20*).

It is difficult to tell exactly how close a friend of Tacitus Pliny was. Certainly only Tacitus receives as many as eleven letters in Pliny's collection of letters from Book I to Book IX. As these letters show, their literary interests were a bond, but other things inevitably threw them together too. They were making their way up the *cursus honorum* at about the same time (see Introduction p. 8). Also, both were lawyers; and they acted for the provincials of Africa as prosecutors of an ex-governor Marius Priscus (II. 11*). This was a long and difficult trial, dragging on for months till AD 100 (S-W, 160); and this professional contact between Tacitus and Pliny may well have led to a development of their friendship. One should perhaps be cautious in claiming for them a shared link with Verginius Rufus, Pliny's guardian. Tacitus, who seems to have admired Verginius Rufus, pronounced the oration at his state funeral in AD 97 (II. 1*); but is this so surprising, since he was consul that year? However, in addition to all the points of contact already mentioned, it might be said that their common experience of being *novi homines* may have drawn them to each other. So perhaps eleven letters to Tacitus are not so surprising.

The letters have led to interesting interpretations. Do they suggest that Pliny wanted the world to think he was a closer friend of Tacitus than he actually was, so that he could bask in the reflected glory of Tacitus' literary reputation? This is to assume (with no supporting evidence) that contemporaries had the same opinion of the relative literary merits of Pliny and Tacitus as many do nowadays. Nor do Pliny's letters necessarily support the view that he felt Tacitus to be his literary superior. Pliny was much given to flattery. When he tells Tacitus 'you are master, I am pupil' (VIII. 7*), he may not believe it; indeed, he may want to give Tacitus encouragement. More likely is that their reputations among contemporaries were on a par (see IX. 23*). And as for their reputation in each other's eyes, there was probably mutual respect, as they read and criticised each other's writings (VIII. 7*).

For the literary and academic side to Pliny's character, see also I. 9, I. 13, VI. 16, VII. 21 ('I am neither reading nor writing — no easy sacrifice') and IX. 6. (Note too VII. 4* on Pliny writing poetry. At fourteen he wrote a Greek tragedy.)

Notes

1-2   **apros tres . . . cepi:** hunting to a Roman did sometimes mean the chase, and some people would pursue an animal across miles of country. But the commonest method was to surround a whole thicket with ropes on which were multi-coloured feathers (*formidines*). No animal would go past this unknown terror: so, once stirred up, a beast would plunge all around the thicket until he got to the place where there was no *formido* — and as he went out here, he would go straight into a huge trap-net. The people stationed at the trap-net would have to be absolutely silent because they wanted the beasts to come straight towards them. This is where Pliny was stationed, apparently on his horse when, in the long intervals of waiting for a beast to emerge, he was wooing Minerva.

2   **ipse:** apparently Pliny organised the hunt and perhaps he actually delivered the death-blow to the beast in the trap-net.

6   **manus vacuas, plenas . . . ceras:** a simple chiasmus, impossible to reproduce in English.

    non est quod: it is just possible to interpret 'quod' as a relative dependent on 'genus'. But cf. Horace: 'non est quod multa loquamur' (*Epistles* II. 1. 30). So translate 'There is no reason why . . . '.

8-10   **iam undique silvae . . . incitamenta sunt:** some of his letters (see below) show that Pliny valued, indeed needed, peace and quiet for reading and writing; and he found this in the country.

9   **venationi datur:** 'is accorded to hunting'.

11   **ut . . . sic etiam:** 'not only . . . but also'.

12-13   **experieris non Dianam . . . Minervam inerrare:** Diana was the goddess of hunting, Minerva the goddess of wisdom and so of literary pursuits. Tacitus apparently took up this reference to the two goddesses, subsequently telling Pliny that Minerva and Diana should be equally worshipped (IX. 10*). From these two letters we gain a rare glimpse of Tacitus and learn that he used to go hunting.

Exploration

'Silvano invicto sacrum. Gaius Tetius Veturius Micianus, praefectus alae Sebosiannae, ob aprum eximiae formae captum, quem multi antecessores eius praedari non potuerunt, voto suscepto libens posuit' (*RIB*

1041). (Sacred to the unconquered Silvanus. Gaius Tetius Veturius
Micianus, prefect of the Sebosian cavalry, put this up of his own free
will having undertaken a vow, because of the capture of a boar of
exceptional beauty, which many people before him were not able to
catch.)
The subject of hunting may be a profitable lead-in. The above inscription (on an altar found near Stanhope, Co. Durham) may be read
and translated by a more advanced group of students; an easier approach
might be to list the words which are to do with hunting, after a preliminary reading.
The Roman method of hunting will need to be explained, and this
may stimulate comparisons with our ideas of 'sportsmanship'. Paoli
(243ff.) states that the origin of hunting as a sport (as distinct from
hunting as a means of obtaining food) is uncertain. The sport probably
arose and continued alongside professional hunting by trained slaves for
food and for protecting flocks.
The letter opens with a hunt, but quickly passes on to the main
theme, that of writing. We see here that Pliny takes his own literary
efforts seriously, and seizes every possible opportunity to escape from
the everyday world in order to pursue them.
Perhaps it should be elicited that Pliny is behaving oddly and in this
letter is trying to get Tacitus to do the same. However, he cannot conceal his exhilaration at having been a success on a boar-hunt.

Questions

1 Why will Tacitus laugh?
2 Is the word 'inertia' (line 3) used disapprovingly?
3 Why did Pliny go on the hunt at all? Did he feel obliged to do
  so? Was it to satisfy the servants at his villa? What would
  happen to the boars?
4 In what sentence does he suggest that time is precious?
5 What are the arguments he gives for writing on a hunt? Do you
  agree with them?
6 What does Pliny mean by 'experieris non Dianam magis monti-
  bus quam Minervam inerrare' (lines 12-13)?
7 Is this letter about hunting or writing?

I. 9

The letter

Pliny draws a sharp contrast between his life in Rome and the days he
spends so idyllically at his villa at Laurentum.

Minicius Fundanus, the recipient of this letter, became *consul suffectus* in AD 107. About seventeen years later he became proconsul of
Asia. He was a friend of Plutarch. Pliny wrote other letters to him: IV.
15*; VI. 6*; and perhaps VII. 12*. In V. 16* Pliny writes feelingly
about the death of Fundanus' younger daughter.

See I. 6 for other letters in this Selection concerning Pliny's literary
interests.

Notes

1-2 **mirum est ... non constet:** as so often, a difficult first sentence.
'quam', meaning 'how', is followed by the three subjunctives 'constet
... videatur ... constet'. 'singulis diebus' is Form C (dative), and
'diebus' has to be supplied again with 'pluribus iunctisque'. The metaphor being used is from accountancy.

3-4 **officio togae virilis:** a family ceremony at which the son of a Roman
citizen exchanged the *toga praetexta* of boyhood for the stripeless *toga
pura*. It took place when a boy was aged between fourteen and sixteen.

4 **sponsalia:** the ceremony of promising to contract a marriage, which
sometimes took place years before the marriage itself.
nuptias: for the Roman ceremony of marriage, in many respects similar
to our own, see Paoli, pp. 116ff. and *CLC* Unit III Stage 29 and
Handbook pp. 123ff.

5 **signandum testamentum:** the signing of a will required the presence of
seven witnesses, who were if possible obliged to be present when the
will was finally opened.

5-6 **in consilium:** a senator with legal experience might be co-opted, as Pliny
was, to the Emperor's *consilium* (IV. 22) to advise on legal decisions.
More often he would join the panel to advise magistrates such as the
praetors or the *praefectus urbi* (J. A. Crook, *Law and life of Rome*,
pp. 87ff.).

8-9 **quot dies quam frigidis rebus absumpsi!:** see Martial, IV. 78 for further
examples of time-consuming, and perhaps pointless, duties in the city,
such as the morning *salutatio* and attending on praetors.

9-10 **in Laurentino meo:** Pliny's very considerable country-house and estate
seventeen miles south of Rome. Pliny describes it in detail in II. 17*.
Pliny was a wealthy man and also had estates at Comum and Tifernum-
on-Tiber (IV. 1, for instance, for his estates at Tifernum-on-Tiber).

11 **cuius fulturis animus sustinetur:** the antecedent of 'cuius' is 'corpori'.
The meaning of this difficult relative clause may be reached by con-
verting it to a causal clause: 'for it is by bodily fitness that the mind is
kept healthy'.

12-13 **apud me:** the meaning may be either 'in my presence' or 'at my house'.
15 **spe:** here means 'ambition' rather than 'hope'.

18 μουσεῖον: (or Museum) had literary associations and was originally the
   name for a place connected with the Muses. The most famous Museum
   was at Alexandria: about 100 scholars worked there and it was a centre
   of scholarship. Pliny wants to convey the idea of his Laurentine villa
   being his centre of scholarship and literature.
20-1 **multum ineptos:** a colloquial usage: 'exceptionally silly'.
21-2 **relinque teque studiis vel otio trade:** compare this, and indeed Pliny's
   whole desire to escape, with I. 3*, where he advises the well-off
   Caninius Rufus from Comum to shut himself away in his retreat with
   his books.
22 **Atilius noster:** Atilius Crescens, a friend of Pliny from Cisalpine Gaul
   (S-W, 107).
23-4 **otiosum esse quam nihil agere:** Pliny summarises his days at Laurentum
   by 'otiosum esse', and his trivial round in the city by 'nihil agere'.

Exploration

'No, Sir, when a man is tired of London he is tired of life' (Dr Johnson
to Boswell).

'I have passed all my days in London, until I have formed as many
and intense local attachments, as any of you mountaineers can have
done with dead nature' (Charles Lamb to Wordsworth).

This letter suggests the age-old discussion of town v. country. Such a
discussion could begin with a reading of Pliny's description of his
Laurentine villa (II. 17*) and a look at the plan of it (for which see
The Letters of the Younger Pliny, Penguin, 305).

After examination of the reasons Pliny gives for disliking life in Rome
and for so warmly praising life at Laurentum, it should become clear
that he is not extolling country against town in broad terms. Nowhere
does this letter suggest that he regrets his public career in Rome; he is
simply weary of the trivial social obligations that fall to someone in his
position, and implies that he regrets spending on them the time which
could be passed in literary pursuits at Laurentum.

As in so many letters, we see Pliny's deep commitment to literature,
which is the central point of his description of Laurentum and indeed
of the letter. He ends the letter with an elegant compliment to his
friend Atilius by quoting his *bon mot* to tie up the two main threads of
thought.

For a poor man's view of life in Rome, and his reasons for retiring
to the country, see Juvenal, *Satires* III.

Questions

1 What are the activities which Pliny says take up a day in
  Rome?
2 Do you think he means it when he says that they are useless?
  Do you agree that they are?

3 What effect on him does he say his activities at Laurentum have?

4 What unpleasant features does he imply that city life has, by mentioning their absence at Laurentum?

5 What do we gather from the letter about the setting of Pliny's villa at Laurentum?

6 Translate the last sentence so that Atilius' saying is expressed 'both profoundly and wittily'.

7 What is this letter about? Do you consider it a serious statement or merely the expression of a mood? Do you think Pliny became equally tired of Laurentum if he stayed there too long?

I. 13

The letter

Pliny approves the large number of *recitationes* but deplores the apathy of people attending them.

Q. Sosius Senecio (*consul ordinarius* in 99 and 107) was son-in-law of Julius Frontinus (*ILS* 1105; 8820) — the governor of Britain immediately before Agricola and an influential friend of Pliny. Senecio was a friend of Calvisius Rufus (for whom see IX. 6), and also of Plutarch who dedicated his *Parallel Lives* and other works to him (see C. P. Jones, 'Sura and Senecio', *JRS* LX, (1970), 98-104, for a persuasive reconstruction of Senecio's career).

See I. 6, for other letters in this Selection concerning Pliny's literary interests.

Notes

1-2 **toto mense Aprili:** in April there were eighteen *ludi* (public holidays), so this was a good month for *recitationes*.

2 **nullus fere dies:** *sc.* 'fuit'.

recitaret: the reading of their own works by writers to an audience was something that belonged to the imperial period. (The Elder Seneca said that Asinius Pollio had started to invite guests to readings of his works in Augustus' time.) Clearly, if Pliny is right, *recitationes* were not taken too seriously by many of those invited. See VI. 15* for a facetious interruption at a *recitatio*. Juvenal and Martial complained of the number of *recitationes*, Juvenal listing among the disadvantages of living in Rome the perpetual dread of fires and falling houses and 'poets reciting in the month of August' ('et Augusto recitantes mense poetas', III. 9). Further reading on *recitationes: OCD* under 'recitationes' (a very useful entry); Carcopino, 214ff.

3 **vigent studia:** Pliny is pleased that literature is flourishing — an in-

direct compliment perhaps to the new freedom of the Emperor Nerva's reign after Domitian's principate, this letter dating from April 97.

4 **ad audiendum**: one of three different usages of gerunds and gerundives contained in this letter. They are:

(i) Purpose: 'ad audiendum' here.

(ii) Form D (genitive) of reference or object:
'tempus audiendi' (5)
'a scribendi recitandique studio' (17-18)
'in audiendi officio' (25)

(iii) Obligation: 'laudandi probandique sunt' (17)

Students may need some revision of gerunds and gerundives in view of these diverse usages in one letter.

coitur: present indicative passive, used impersonally.

5 **tempusque audiendi . . .** : 'and they waste with gossip the time when they are supposed to be listening': 'fabulis' has nothing to do with 'audiendi'.

11 **Claudium Caesarem ferunt**: see Suetonius, *Claudius* 41, for Claudius giving his own *recitationes*.

13 **Nonianum**: the consul M. Servilius Nonianus, an orator and historian (Tacitus, *Annals* XIV. 19, etc.).

13-14 **subitum recitanti . . .** : 'suddenly and unexpectedly came to hear his recitation'.

19-20 **erant sane . . .** : 'admittedly most of them were friends'. Pliny claims to be acquainted with most of the writers of his day.

20-1 **qui studia . . . amet**: an ellipse of what is in any case an idiomatic use of the consecutive clause. Expanded, it might read 'qui studia ita amat ut non simul et nos amet'.

22 **repetere**: 'seek again'.

23-6 **quod non recitem . . .** : 'but not something to recite, lest I may seem to have been not a hearer but a lender to those people whose recitations I have attended. For as in other things, so in the service of being a listener, appreciation ceases to be so when it is asked in return.'

Exploration

The mass of people who attend *recitationes* are ill behaved and have scant interest in literature. Pliny develops this theme, giving a colourful picture of a typical *recitatio* (4-10). He contrasts the apathy of his contemporaries with the enthusiasm of the Emperor Claudius, who on one occasion came uninvited to a *recitatio*, and with his own conscientious attendance at these occasions, even to the point of staying in Rome longer than he has intended.

This last point invites comparison with I. 9: 'quot dies quam frigidis rebus absumpsi!' he exclaims of the daily round of social events — but the present letter makes it clear that *recitationes* come much higher in his estimation.

Of Pliny's sincere interest in literature there can be no doubt, and his publication of this letter must be meant to commend *recitationes* and censure those who treated them frivolously. Nonetheless, his protestations ring slightly false: does not the letter betray his real feeling

that attendance at *recitationes* is a social duty that comes very close
to boredom? A contrary view might be that he does enjoy these occasions, but is remarkably smug about saying so. What would Senecio
feel when he read this letter on his remote German frontier?

Questions

1 Were the *recitationes* getting good audiences?
2 What was the behaviour of those who went to hear them?
3 Was Claudius 'otiosissimus'? How did he put to shame those
   who were so?
4 Explain 'queritur se diem (quia non perdidit) perdidisse'
   (line 16).
5 In what two words does Pliny summarise the attitude of the
   audiences?
6 What claim does Pliny make for himself?
7 Explain the last two sentences in your own words. What is
   Pliny afraid people will think, if he gives a recitation?
8 Does Pliny anywhere in this letter mention the possibility that
   the *recitationes* might be bad?
9 Did Claudius arrive because there was a recitation or because it
   was Nonianus?
10 Does Pliny give the impression that anyone enjoyed the
   recitations — or that he himself did?
11 Does he think that appreciating art is a duty rather than a
   pleasure?
12 Are the words 'desidia aut superbia' (line 18) fair?
13 Do you find this letter irritating? Pompous?

I. 19

The letter

Pliny offers to a fellow townsman of Comum the extra money which he
needs to become a knight.
  There are in this Selection six other letters which directly concern
patronage: II. 6 (the graded dinner-party); VI. 23 (he asks that
Cremutius Ruso be allowed to assist him in a case); VI. 25 (Pliny has
given 40,000 sesterces to a fellow-townsman); and three letters dealing
with the patronage of communities: IV. 1 (the dedication of a new
temple at Tifernum-on-Tiber); III. 6 (a statue for some public position
in Comum); IX. 39 (the rebuilding of the Temple of Ceres on his

estates). Among numerous other letters dealing with patronage, see for instance the long and interesting II. 13*.

For a discussion of patronage, see pp. 20-1.

Romatius Firmus is known only from this letter and IV. 29*. This letter shows that he came from Comum and was much the same age as Pliny.

Notes

1 **municeps tu meus**: numerous letters show Pliny's affections for Comum and his concern for its people (e.g. II. 8; IV. 13*, a school-master for Comum; VI. 25).

1-3 The verb 'to be' is omitted at least twice in the first sentence.
ab ineunte aetate . . . familiaris: it was a process amounting to patronage to advance one's own friends and friends of the family. See, also, II. 9 for Pliny as a political patron.

5-6 **quod apud nos decurio es**: this noun-clause is the subject of the verb 'indicat'. 'quod' may be translated as 'the fact that'.

6 **decurio**: this sentence is the only evidence we have that Italian municipal councillors needed a property qualification of 100,000 sesterces.

7-8 **ad implendas equestres facultates**: in order to become an *eques* a person had to have, certainly since late Republican times, a minimum of 400,000 sesterces. For a full discussion on the benefits of equestrian status, see S-W, 130-1.

8 **te memorem**: *sc.* 'fore'.

9-10 **ego ne illud quidem**: 'I do not even give that reminder which I would feel it my duty to give if I did not know . . . '

12 **ut**: here the word almost has the meaning of 'because'.
utare: present subjunctive, second person singular.

Exploration

Some explanation will be needed before students translate this letter. First it should be established that Romatius Firmus came from Comum (this will probably be elicited from the first three words of the letter). Then on to the equestrian qualification. In answer to the inevitable question, 'How much was 400,000 sesterces?' a group can be given other valuations for comparison:

1,200 sesterces = a legionary's annual pay
12,000 sesterces = the expenses of a Byzantine ambassador going to Rome (X. 43*)
3,318,000 sesterces = money spent on an uncompleted aqueduct (X. 37)

Why did Pliny choose to publish this letter? It is a guess that the letter was substantially altered by Pliny for publication. Was it likely that he would remind Firmus that they were fellow-townsmen and went to school together, or that Firmus' capital was only 100,000 sesterces before Pliny's gift of three times that amount? This kind of information

might have been put in to provide the general reader with details of
Pliny's connection with Firmus. By publishing this letter, Pliny was
surely seeking the approval of his contemporaries.

Questions

1 What does Pliny give as the reasons for the kindness he is
 going to bestow?
2 How much is Pliny giving? Is it a large sum? 'ego ne illud
 quidem admoneo . . . ' (lines 9ff.): what is the self-contradiction
 which follows?
3 What phrases in the letter would lead you to believe that
 Pliny is writing to someone of his own age, if he had not said
 so?
4 Will the gift bring any benefit to Pliny himself? Why is he
 giving it?
5 Do you think that the letter is tactful? What would be your
 reaction on receiving it?

II.6

The letter

Pliny has been to a dinner-party where the host graded the food and
wine according to his guests' social status. He warns his young friend
Avitus to avoid such insulting practices.

The recipient is perhaps Junius Avitus, whose premature death is
recorded in VIII. 23*.

For other letters in this Selection which deal with patronage, see
I. 19.

Notes

1-2 **quemadmodum acciderit:** this indirect question is dependent on both
 'altius repetere' and 'refert'.
2 **homo minime familiaris:** refers to Pliny himself, the subject of 'cena-
 rem'. It may help if students supply 'ego' before this phrase.
8 **nam gradatim amicos habet:** it is clear that the host is discriminating
 not according to intimacy but by social status because Pliny himself,
 though 'minime familiaris' (line 2), is receiving the best food.
9 **animadvertit qui:** an antecedent to 'qui' has to be supplied.
12 **notam:** the mark which the censors put against the name of a man
 whom they were expelling from the senate: thus the word is used
 figuratively to convey any act of degradation.
13 **exaequo:** *sc.* 'eos'.

17 **hercule:** this mild oath marks the switch from narration to a direct
address to Avitus.

18 **quo utaris ipse:** this phrase (which, again, needs an antecedent supplied) can be postponed till after 'communicare cum pluribus'.
illa: refers to 'gula'.

19-21 **si sumptibus ... consulas:** 'if you are to cut down your expenses, and
you may look after them rather more honourably by continence on
your part than by contumely to others'. 'parcas' and 'consulas' are
potential subjunctives. Note also the word-play on 'continentia' and
'contumelia'.

27 **cum:** 'while' or 'although'.

28 **turpius iunguntur:** 'are even more shameful when combined'.

Exploration

It is not easy to say how common such graded dinner-parties were, but
they are certainly alluded to by both Martial (III. 60) and Juvenal
(*Satires*, V), who describe in detail the vile, even rotten, food which is
placed before the *clientes* in contrast to what the *patronus* eats. Juvenal
adds that it is done as a deliberate insult: 'hoc agit ut doleas; nam quae
comoedia, mimus quis melior plorante gula?' (157-8) But he goes on to
add that in treating the guests like this, the *patronus* is showing his
wisdom — 'for if you can endure all these insults, you deserve them'.

Of the eight letters in this Selection which bear directly on patronage, this is the only one which shows the custom in an unfavourable
light; but then the others concern Pliny himself as patron, and so it is
not surprising that they should present things more attractively. Even
so, Pliny does seem to have been more than usually generous in his
treatment of slaves, freedmen and other social inferiors (see letters
listed in the notes on V. 19). For a similarly enlightened attitude a
generation earlier, see Seneca, *Letters*, XLVII, 14ff.

It is indeed possible that, like the Seneca epistle, this letter of Pliny
is not really a letter at all but a *sermo* dressed up as a letter, and the
conversation at the dinner-party is a literary device to introduce it.
Students may find this interesting as a possibility, but it is not veri-
fiable — indeed to assume such a thing would be to beg a huge question
which must remain unanswered. For a discussion on the authenticity of
Pliny's Letters as correspondence, see S-W, 11-20.

Questions

1 How would you describe the three grades of guests?
2 What sentence suggests that the procedure was unusual?
3 How does Pliny summarise his own attitude to guests?
4 How does he economise?
5 What metaphor does he use in lines 18-19? Do you find it
strange?

6   What is the play on words in lines 20-1? Try to give an apt translation.

7   Why is Pliny telling his young friend all about this?

8   How would you translate 'luxuriae et sordium' (line 26) to make it clear that they are opposites?

9   Why, in your view, does Pliny call himself 'homo minime familiaris' (line 2)?

10 Of what social position was the man 'qui proximus recumbebat' (line 9)? Were the guests really *amici* or would you use another word? In that case, what would be Pliny's status at such a dinner?

11 Summarise what Pliny is really objecting to. Do you agree with him?

12 Is his attitude consistent with the previous letter? Or with others you have read?

13 Does Pliny usually treat patronage in such an unfavourable light?

14 What other letters give a different picture? Why?

II.8

The letter

Work keeps Pliny at Rome and he pines to be at Comum.

Caninius Rufus (to whom, among other letters, IX. 33 was also written) was well-off and came from Comum, where he had a house that Pliny much admired (I. 3*).

Private citizens of Comum crop up regularly in Pliny's letters. Calvisius Rufus, to whom he wrote IX. 6, was a town-councillor of Comum; so was Romatius Firmus for whom he made up the property qualification to become an *eques* (I. 19). Metilius Crispus, to whom Pliny gave 40,000 sesterces on his becoming a centurion, was also from Comum (VI. 25). Clearly his roots meant much to Pliny.

Notes

2   **ad Larium nostrum:** Lake Como has, in post-Roman times, taken its name from the town of Comum (the modern Como), but the Romans called it Larius. (See Virgil, *Georgics*, II, 159.) Students will need the location of Comum established — by its lake, just south of the Alps, on the border of modern Italy and Switzerland. Some of them may have passed through the town by train or car and will be able to describe how it appears today. What will students gather from the word 'nostrum'? It shows that Pliny and Caninius were friends and that both considered Comum to be their home.

3 **altissimus:** probably 'deep' but the meaning could possibly be 'high'.

6-7 **ut aegri vinum balinea fontes:** see VII. 21, where Pliny when ill takes a bath and some wine.

7-8 **numquamne . . . abrumpam?:** this is not the only letter in which Pliny pines to escape from Rome. See I. 9, where he longs for the seclusion of his villa at Laurentum: but see IX. 15, where Pliny, on his Tuscan estates, wants to be kept in touch with what is happening in the city.

Exploration

It can never be known how much Pliny revised each letter for publication, but the charm of this one seems to lie in its spontaneity.

Their careers took men like Pliny away from the places where they had grown up, perhaps for all their lives. The life of a senatorial centred on Rome, with the expectation of appointments in the provinces (see Introduction, I: Background to Pliny's career, pp. 3-6). If time permits this letter can be read side by side with Catullus' poem 'paene insularum' (XXXI in *CLC* Unit IV) for a comparison between Pliny's nostalgia for Comum and Catullus' joy at being home again at Sirmio after his post in Bithynia. Are nostalgia and the joys of home better expressed in poetry than in prose? It would be worth asking a group to think of parallel examples from English literature. Again if time permits, I. 3* could be read in translation — where Pliny talks to Caninius of his (Caninius') house and of Comum generally.

Questions

1 Where is Caninius?
2 What are Pliny's feelings at the time he is sending the letter?
3 What is preventing him from following his own inclinations at the moment?
4 Which is faster — the rate the work is piling up, or the rate he is disposing of it?
5 Find all the similes in this letter, and then all the metaphors.
6 Do you think Pliny means what he says?
7 Is 'escape to the country' a common literary convention?

II.9

The letter

Pliny is anxious about Sextus Erucius' candidature for the tribunate, and asks Apollinaris to give him his support.

For seven letters in this Selection concerned with patronage, see I. 19. Domitius Apollinaris, to whom Pliny wrote this letter, was *consul suffectus* in AD 97 (the same year as Tacitus) and was perhaps the poet Martial's patron (S-W, 156-7). Date of this letter: 'Either late 97 or late 100 is arguable' (S-W, 156). So Pliny is probably writing to an ex-consul.

Notes

1 **petitio Sexti Eruci mei:** Pliny comes straight to the point: his friend Sextus Erucius is standing for tribune of the people (*tribunus plebis*) and Pliny is concerned that he should be elected. (The election would take place in the senate, and the *comitia centuriata* would formally confirm the senate's vote: S-W, 157.)

See the first section of the Introduction, Background to Pliny's career (pp. 5-6), for a description of the typical career-patterns of senatorials in the first century AD. In a senatorial's promotion a man became either tribune of the people or aedile between being quaestor and praetor. Sextus Erucius had already been quaestor (line 5). Since Sulla's reforms (mostly 81 BC) a quaestor had automatically become a member of the senate. Sulla had raised the number of quaestors to twenty, and this was still the number in Pliny's day; the number of tribunes also had remained unaltered at ten. In Sulla's time there were eight praetors, but by the time of this letter there were eighteen. Sulla had fixed the minimum age-limit for being quaestor to thirty, praetor to thirty-nine and consul to forty-two; but there were lower age-limits operating in Pliny's time (see, for instance, S-W, 73ff). See Syme, *Tacitus*, I, 64ff on the senatorial career in Tacitus' and Pliny's time.

Senatorial promotion was very much a pyramid rather than a ladder. The number of people who reached the top was small; and even this early in Erucius' career Pliny is needing to drum up all the support he can. Erucius' own career exemplifies the hazards in a senatorial's promotion. It may be presumed that he was elected *tribunus plebis* — would Pliny otherwise have published the letter? But he did not become consul for the first time till AD 117. He had to wait till AD 146 for his second consulship and the office of *praefectus urbi*, under the Emperor Antoninus Pius. He died the same year (S-W, 157; Syme, *Tacitus*, I. 242; and II. 477). It seems reasonable to suppose that his political career under Hadrian was adversely affected by his uncle Septicius Clarus' disgrace in *circa* AD 122 (see below, lines 15-16).

2-3 **adficor cura . . . patior:** students may need help with this sentence — for instance, that 'cura' is Form E (ablative), and that 'sollicitudinem' is the antecedent of 'quam'.

4-5 **ego Sexto latum clavum a Caesare nostro, ego quaesturam impetravi:** the quaestorship automatically gained a man admittance to the senate; and senators had the *latus clavus*. But this sentence seems to be distinguishing between the grant of the *latus clavus* which Pliny obtained for Sextus Erucius and Erucius' quaestorship. It is likely that from Augustus onwards ambitious young men not of senatorial families but aspiring to the senate applied to the emperor for special leave to wear the *latus clavus*. See A. H. M. Jones, *Studies in Roman government and law*, 31-2.

Of the twenty quaestors elected each year, two became quaestors of the emperor (*quaestores Caesaris*), two became urban quaestors and four quaestors to the consuls; the rest drew lots for the public provinces (S-W, 292). Pliny and Calestrius Tiro were *quaestores Caesaris* together (VII. 16); and Rosianus Geminus, the recipient of VII. 24, was Pliny's *quaestor consularis* in AD 100. The *quaestores consulares* of the *consules ordinarii* served under the *suffecti* too, for the whole year (S-W, 476).

9 **quae causa:** these words are part of the conditional clause: 'even if this cause did not arouse my enthusiasm . . . '.

10 **adiutum . . . cuperem:** the use of the past participle gives an objectivity to the statement, i.e. 'I would wish him supported'.

12-13 **nam pater ei Erucius Clarus:** 'est' (not 'erat') must be supplied, because 'defendit' must be the present tense. If the father were dead, the word would be 'defendebat'. Erucius Clarus was apparently an equestrian. It is presumably to him, and not his son (the central figure of this letter), that Pliny wrote I. 16*.

15 **defendit:** 'he pleads'. The word implies that Clarus chose to plead defence cases rather than to conduct prosecutions.

15-16 **habet avunculum C. Septicium:** C. Septicius Clarus was an equestrian who progressed to becoming Praefect of the Praetorians in *circa* AD 119. However, Hadrian sacked him, along with others (including his *ab epistulis* Suetonius) in *circa* AD 122 (S-W, 159; Syme, *Tacitus*, I. 87 and note; II. 501; II. 779). Pliny dedicated his *Letters* to Septicius (I. 1*) and Suetonius dedicated his *The Twelve Caesars* to him (S-W, 127).

16-17 **nihil verius . . . :** for this use of 'nihil' in reference to a person, see IV. 22, lines 9-10: 'quo viro nihil firmius nihil verius'.

20 **quantumque . . . valeam:** this is an indirect question, dependent on 'experior'.

22 **tanti:** 'worthwhile' — Form D (genitive) of value.

24-5 **qui quod tu velis cupiant:** 'who will be at your service'. Apparently this is a regular phrase: Horace politely says to the bore, 'cupio omnia quae vis' as an attempt to preserve a formal atmosphere (*Satires*, I. 9. 5). Pliny, however, is using it as a compliment to Apollinaris, the nicety of the phrase being the play on 'velle' and 'cupere'.

Exploration

'anxium me et inquietum habet petitio Sexti Eruci mei.' This opening sentence clearly and economically states the two interrelated themes of this letter, Erucius' candidature for *tribunus plebis* and Pliny's concern that he should be elected. Like other letters, this one gives a fascinating glimpse into the workings of Roman patronage and supplements the cold records of patronage provided by inscriptions. Pliny gives his readers details which Apollinaris probably knew already: that he had started Erucius on his senatorial career, twice recommending him to

the Emperor (lines 4-5). Now he is soliciting the support of Apollinaris,
probably already an ex-consul (see above) and so a man of consequence.
Pliny extols the merits of Erucius and of other members of his family.
Nowadays political candidates seek the support of voters by offering
an attractive programme of legislation. Politicians in the days of the
Republic had done this (for instance, in his bid for the consulship of
70 BC, Pompey had promised to restore the powers of the tribunes;
and after he was elected, he did so). But under the emperors, character
rather than political programme, was more appropriate: after all, policy
was the emperor's concern.

Why was Pliny concerned that Sextus Erucius should be elected?
He speaks of his own *dignitas* being called into question (line 4). *Dignitas* was all-important to a leading Roman and Pliny might well be afraid
for his political standing if he were thought to have deceived the Emperor (line 7). The dating of this letter is relevant: if it was written late
in AD 97 (S-W, 156), Pliny had not yet been consul and he would be
very concerned that nothing should jeopardise his attaining the office
early, before he was forty. His hopes were in fact rewarded by a suffect
consulship in AD 100; but perhaps the inversion of the first sentence
does betray Pliny's priorities, his own career first and Sextus Erucius'
career second.

With the possible exception of X. 96, this is the most difficult of
the letters in the selection, for two reasons: (i) the complexity of some
of the Latin; and (ii) the bulk of background knowledge that is required. For these two reasons it is suggested that a good deal of careful
groundwork be done before the letter is even prepared, let alone translated. The background material should be tackled first. (For this, see
Introduction, particularly I: Background to Pliny's career (pp. 1-7) and
the notes above.) To give too much information at this stage can be
counter-productive. More can be given, after the letter has been translated.

Next to the language. One problem here is the way Pliny festoons
people with strings of complimentary adjectives ('iuvenem . . . dignis-
simum', lines 10-12; 'avunculum . . . nihil fidelius novi', lines 15-17).
He does something similar with verbs in lines 19ff. Then there are
connecting relatives (e.g. 'quae causa', line 9). When it comes to translating the letter, it can comfortably be taken in two parts, the division
coming in line 12.

After the letter has been translated, further exposition and discussion
will be necessary. (*CLC Handbook* for Unit III has some useful suggestions for class activity in connection with patronage, p. 112.) Others of
Pliny's letters could be brought into the discussion. (Some could be
read in translation.)

It would seem desirable, after all the time and trouble taken over
this letter, to reinforce its importance by getting a group to do some

written work on patronage (and perhaps not just on patronage in
Roman times).

Questions

1 What does 'mei' (line 1) imply?
2 What is Pliny's chief fear?
3 What office is Sextus Erucius seeking?
4 Who will vote for the filling of the office?
5 What does Pliny say was his attitude when he himself was a
  candidate? Do you believe him?
6 What profession does the young man's father follow?
7 What is Pliny's relationship to the family?
8 What does Pliny say he himself is doing to support the young
  man?
9 What does he ask Apollinaris?
10 Do you think Pliny would be glad of a duplicating machine
  for this letter? Or do you think it was the only one of its kind
  he was writing?
11 Is Pliny being kind to the young man? Or is he doing it for
  selfish motives?

III.6

The letter

Pliny has bought a bronze statue and, wishing to present it to his home
town of Comum, he asks Annius to have a pedestal made for it.

For the seven letters in this Selection dealing directly with patronage,
see I. 19. This present letter and two others (IV. 1; IX. 39) concern
Pliny as a public benefactor. Two other letters (II. 16*; V. 1*) were
written by Pliny to Annius Severus, who was his agent.

Notes

1-2 **Corinthium signum:** Corinthian metalware had for some time been
  very much the fashion. Trimalchio (Petronius, *Satyricon*, 50) mocked
  this fashion, explaining that what he had was genuine Corinthian
  because it was made by Corinthus.
    Sherwin-White believes that the statue was of an old woman
  (S-W, 226): this was a favourite subject in late Hellenistic sculpture.
  If this is true, it is odd of Pliny to use 'senem' (line 6) without further
  qualification, though 'papillae iacent' (line 9) could be taken to suggest
  femininity. Students will have to come down on one side or the other,
  or else translate 'senem' as 'old person'.

2 **festivum:** this word is often used as a vague term of approbation.
7 **ut spirantis:** 'as if belonging to a living person'.
9-10 **a tergo . . . ut a tergo:** the idea is, 'From the back view also it shows the same age, so far as is possible from a back view.' For other examples of this usage, see Lewis & Short under 'ut' I.B.4.b.
15 **in patria nostra:** i.e. at Comum. For Pliny's affection for his home town of Comum and his interest in the people there, see II. 8 and notes.
16 **in Iovis templo:** presumably the principal temple.
19 **iam nunc:** both words should be translated.
   marmore: since Pliny wants his name, and possibly his honours, inscribed, marble is the obvious material for the pedestal.
19-20 **quae nomen meum honoresque capiat:** inscriptions reveal how common it was for a patron's name and his career to be set up in some public place (for the career of Pliny himself, see Appendix no. 1). Which *honores* are meant? Perhaps in particular the consulship, Sherwin-White dating this letter to after Pliny had been consul and perhaps to after he had accepted the curatorship of the Tiber (S-W, 225).
25 **ad paucos dies:** this refers to the length of his future stay.
25-6 **neque enim diutius abesse . . . :** *sc.* 'patiuntur'.

Exploration

A group could well start off its reading of Pliny with this letter: it sticks vividly in the memory and both the Latin and, at a superficial level, the subject-matter are straightforward.

Discussion may first turn to Pliny's views on sculpture. He presents the statue's trueness to life as the only evidence of its merit, thus expressing a view which has not gone unchallenged elsewhere but which was prevalent in his own time. It was this attitude which had led Greek sculptors into the worst excesses of ultra-realism which we find in, for example, the famous Laocoön sculpture, where the central figure, wrestling with the serpent, has tautened every muscle and sinew in his body: the work is a *tour de force* of craftsmanship and almost a stone text-book of human anatomy, but the effect is gross, violent and a far cry from the calm elegance of the classical period. For a picture of the sculpture, see for example plate 194 in John Boardman, *Greek art*, rev. edn (Thames and Hudson, 1973).

Pliny's description of his statue suggests something of this type: apart from the details which he gives, the very subject — a nude aged figure with its wrinkles and veins — was a popular choice for sculptors who wished to display their mechanical expertise. It might be useful for students to look at pictures, first of this kind of sculpture, and then by comparison of figures from the classical Greek period. Is realism the sole criterion of merit, as Pliny implies, or are there more important considerations? Finally, students should be shown examples of native Roman sculpture: the Roman sculptors excelled in portraiture, and as good an example as any is the Ara Pacis in Rome (13 BC), where the realism concentrates not on anatomy but on the facial expressions (for the Ara Pacis, see Sir Mortimer Wheeler, *Roman art and architecture*, 163ff). Slides nos. 51-63 in the *CLC* Unit IV collection provide further examples of portraiture.

Next, students could express their reactions to lines 19-21, where Pliny wants his name and honours put on the statue's pedestal. This was common Roman practice, as any number of inscriptions show. Different periods of history have different standards of behaviour and the Romans were much less inhibited than we are, much franker in their vanities. Why does Pliny want his honours recorded here? For one thing, because it was accepted practice. For another, as the donor of the statue, he should get something in return for his gift, in this case enhanced *dignitas* and a chance to be remembered even after he was dead (a Roman's idea of immortality). Perhaps, too, since he had probably been consul when he gave this statue, Comum could have the reflected glory of its patron's consulship. These things taken into account, Pliny may be seen in a more favourable light.

Pliny's benefactions were considerable (see Appendix no. 1). As a tail-piece to the discussion, students could list these benefactions, noting that this inscription in the Appendix (which lists Pliny's honours as well) was posthumous.

The final four lines of this letter ('destino', line 23 to the end) raise a more complex issue. Is there evidence here of Pliny having amalgamated two letters? 'destino' does not equal 'polliceor' (line 24), and perhaps 'gaudes' (line 24) onwards is part of a second letter, where Pliny has heard from Annius and is able to write back and say that he is definitely coming to Comum but not (as Annius hoped in his letter?) for a prolonged stay ('gaudes quod . . . ', line 24, reads remarkably like the openings of three Pliny letters — IV. 16; VI. 26*; VII. 23*). An opposing view represents this shift from 'destino' to 'polliceor' as Pliny being artificially spontaneous. In the final analysis it is impossible to say which (if either) of these views is correct; but a discussion of them focuses attention on the fact that the letters were revised for publication.

Questions

1 What is Pliny's estimate of his own artistic judgement? Does it sound sincere?

2 What does he say are the advantages of the statue being a nude?

3 What criteria is he using to estimate the statue's worth, or at least his own approval of it?

4 Where does Pliny want to put the statue? Do you think his motives are religious ones?

5 What two choices does Pliny leave to Annius? Are they real choices?

6 What tenses are 'gaudes' (line 24) 'contrahes' (25) and 'adiecero' (25)?

7 What phrase makes it clear that Annius was Pliny's agent? What sentence suggests that the two men were on good terms?

8 What is more important to Pliny — that he should set up the statue in his home town or that he should dedicate it to Jupiter? Would he have understood what we meant if we had been able to ask him whether his motives were religious ones?

9 What do you think was keeping Pliny in Rome? See I. 9.

III.14

The letter

Pliny tells of the violent death of Larcius Macedo and of an earlier incident which seemed to foreshadow it.

No definite details are known about Acilius, the recipient, but see S-W, 246.

Notes

1-4 **rem atrocem . . . meminisset:** a possible approach to this difficult sentence might be to take first the six words, 'rem atrocem Larcius Macedo passus est', to translate them, noting the inversion, and then to build the rest of the sentence around them. Thus 'nec tantum epistula dignam' is seen as adjectival to the object 'rem'; 'a servis suis' is adverbial to 'passus est'; and the rest of the sentence is adjectival to the subject 'Larcius Macedo'. (A blackboard demonstration of this could be helpful.) The words 'superbus . . . meminisset' are most easily translated as a separate English sentence beginning, 'He was . . .'

The murder by slaves of important Romans was apparently not common: but in AD 61 the Prefect of the City, Pedanius Secundus, had been killed by one of his slaves (Tacitus, *Annals*, XIV. 42-5).

2-3 **superbus alioqui dominus et saevus:** Pliny would not approve, in view of his *humanitas* towards his own slaves and freedmen (II. 6; V. 19; VIII. 16).

3 **servisse patrem suum:** how did Larcius Macedo, the son of a slave, become *vir praetorius* (line 2)? Sherwin-White suggests that 'possibly this man is the adopted son of a household favourite' of the senatorial Larcii (S-W, 247). No other case is known this early of a senator being the son of a slave. In the Julio-Claudian period Curtius Rufus, *legatus* of Upper Germany in AD 47 and later proconsul of Africa, was reputedly the son of a gladiator. Tacitus commented, with apparent relish, 'I would not wish to speak falsely and I am ashamed to speak the truth about the origins of Curtius Rufus, who some have claimed had a gladiator for a father' (*Annals*, XI. 21).

4 **meminisset**: generic subjunctive, 'qui . . . meminisset', meaning 'the kind of man who remembered . . . '

5 **circumsistunt**: Pliny slips into the 'vivid present' tense, a common Roman (and colloquial English) habit when telling a story.

7 **verenda**: not the commonest word, but one used by Pliny's uncle (see Lewis & Short under 'vereor').

8 **fervens pavimentum**: almost certainly not the hot courtyard but the floor of the *caldarium*. The word 'effertur' (line 11) suggests Macedo's body being carried out of a building. As a praetorian he would be rich enough to have this kind of elaborate bath-house.

11 **quasi aestu solutus**: Juvenal (*Satires*, I. 142-4) talks of over-eating causing death in the baths.
effertur: the usual word for carrying out a body for burial. Perhaps Pliny uses it here to add colour to the story.

12 **concubinae**: not a common feature of upper-class Roman households. The mention of them by Pliny can only imply disapproval of Macedo. Compare Tacitus' mention of them in connection with Tigellinus, of whom he disapproves (*Histories*, I. 72).

17-18 **ita vivus vindicatus, ut occisi solent**: going right back into Republican times, it had been the custom (formalised into law by Sulla) that a man's household of slaves (*familia*) should all be killed, if the master was murdered by one or more of his slaves. When Pedanius Secundus had been murdered by a slave, his whole household of 400 slaves were eventually executed, but only   after considerable public debate (S-W, 463). Pliny seems to imply that Macedo's death was a foregone conclusion: hence the execution for murder of those slaves who were rounded up.

23-4 **dies feriatus**: no public business was allowed on such a day.

24-5 **addam quod . . . succurrit**: 'mihi' must be supplied to make this sentence clear.

25 **cum in publico Romae lavaretur**: almost a complete cross-section of Roman society visited the public baths. For instance, Suetonius recalls that the Emperor Titus used to go to the public baths (*Divus Titus*, 8).

Exploration

With its immediate appeal, this letter is bound to stick in the memory and could usefully be made the springboard for as much discussion as possible.

First, there is Pliny's attitude to the murder of Macedo. The point of the letter seems to be reached in lines 18-21 ('vides quot periculis . . . perimuntur') where Pliny expresses concern at the risks run by masters, even lenient ones, of violence from their slaves. However, in the first sentence of the letter he has the horror of the attack on Macedo nicely balanced with the suggestion that, as an overbearing, cruel, master with a servile past, he had it coming to him anyway. So where do Pliny's sympathies lie in this particular case? He carefully does not tell us. The law acted savagely against a household of slaves if one or more of them had killed their master (see Notes, lines 17-18). The justification for such a law is worth debating.

The second paragraph narrates an earlier incident which Pliny says was an omen of Macedo's death. Is this evidence of Pliny's superstition (see IX. 39, where he wishes to act 'on the advice of soothsayers') or is he pandering to the popular liking for omens and significant coincidences?

From this arises a final question: why did Pliny write this account of Macedo's death? This letter seems unlike many of those which savour of spontaneous correspondence. Was it first written for inclusion elsewhere, like the account of the death of his uncle (VI. 16) and that of the dolphin at Hippo (IX. 33)? The first sentence suggests this: Macedo's death was 'a hideous thing and worthy of something more than a letter'. The missing clue is the identity of Acilius, the recipient of the letter. Is he, like Caninius and Tacitus, a writer whom Pliny supplies with incidents for inclusion in his works?

Questions

1 What does Pliny diagnose as the cause of Macedo's unpleasant behaviour towards his slaves?
2 Why did the slaves throw him out onto the hot floor? How would this serve the purpose they intended?
3 How does Macedo's household differ from that of a normal Roman?
4 Why did the slaves run away?
5 How did Macedo get his revenge? Was it only the guilty that suffered?
6 Why does Pliny say he has the opportunity to write more?
7 Explain how Macedo's death had, according to Pliny, been foreshadowed by an omen. Do you think Pliny is serious? How do you view it?
8 Would you say Pliny's sympathies are more with Macedo or with the slaves?
9 Do you find his conclusions about the dangers from slaves pessimistic? Do they follow logically from the story he has just told?

IV . I

The letter

Pliny writes to his wife's grandfather: they are going to pay him a visit, but must stop on the way to attend the dedication of a temple at Tifernum-on-Tiber.

For the seven letters in this Selection concerned with patronage, see I. 19.

L. Calpurnius Fabatus was the grandfather of Pliny's third wife, Calpurnia. Like Pliny, he came from Comum where he was a wealthy citizen and town magistrate. Fabatus' equestrian military career (for which see Appendix no. 3) came prematurely to an end, when he was implicated in the charges against L. Junius Silanus towards the end of Nero's reign (informers had been produced to forge against Silanus' aunt a tale of magical rites and incest with her nephew). Fabatus and others appealed to the Emperor and escaped condemnation (Tacitus, *Annals*, XVI. 8). But the case apparently finished Fabatus' career. Pliny and Fabatus had dealings in connection with each other's properties (see, for instance, VI. 30*). Fabatus is also the recipient of VII. 16 and seven other letters not in this Selection. He died at a great age while Pliny and Calpurnia were in Bithynia, i.e. *circa* AD 112 (X. 120*). Pliny's letters to Fabatus suggest a comfortable, easy relationship, intimate enough for Fabatus to be frank in his criticisms of Pliny (see, for instance, VI. 12*, from which, *inter alia*, Sherwin-White concludes that the old man was sharp and irascible).

Calpurnius Fabatus had a daughter called Calpurnia Hispulla (IV. 19*; VIII. 11*; X. 120*). Hispulla was also the name of the wife of Pliny's early benefactor, Corellius Rufus (see Introduction, pp. 5 and 20). Syme has therefore suggested that the families of Fabatus and Corellius Rufus may have been linked by marriage (*Tacitus*, I. 86).

Notes

2 **mutuo:** there is no word in English corresponding to this enviably brief adverb which in one word expresses 'and the feeling is returned'.

3 **quodam:** this qualifies two rather extravagant words, and could be translated as 'almost'.

4 **differemus:** an illogical word with 'desiderium' as the object. A possible translation is 'which we shall not leave unfulfilled any longer'.

9 **praediis nostris:** Pliny had a considerable investment in land. As well as these estates at Tifernum-on-Tiber, he had a villa at Laurentum sixteen miles south-east of Rome (I. 9) and estates at his native Comum (VII. 11*, for instance). Pliny's farms at Tifernum-on-Tiber alone brought him in more than 400,000 sesterces a year (X. 8*).

9-10 **nomen Tiferni Tiberini:** Tifernum-on-Tiber was over 150 miles north of Rome (X. 8*).

10 **me paene adhuc puerum patronum cooptavit:** Sherwin-White suggests that, whereas Pliny's estates at Comum were inherited from his father, these at Tifernum-on-Tiber may have come to him from his uncle, Pliny the Elder (S-W, 265).

From III. 4* it appears that Pliny was also patron of the province of Hispania Baetica. Emissaries from that province begged him to prosecute an ex-governor, Caecilius Classicus, in Pliny's words, 'implorantes fidem meam . . . adlegantes patrocini foedus'.

There is ample inscriptional evidence of cities and communities 'co-opting' individuals as their patrons and of these individuals accepting the places concerned as clients. So, from Augustus' reign, a community on the larger of the Balearic Islands 'M. Atilium M.f. Gal. Vernum patronum cooptaverunt . . . ' (*ILS* 6098).

14 **templum pecunia mea exstruxi:** two other letters also refer to this
temple (X. 8*, where Pliny explains to the Emperor Trajan his plans
for the temple and asks his permission to put Trajan's statue in it —
this letter dates from AD 98 or 99 — and III. 4*, where he mentions
laying the temple's foundation-stone). This temple was built on public
land, so Pliny had had to ask the town council to allocate him a site
(X. 8*), whereas the temple to Ceres which Pliny proposed improving
(IX. 39) was on his own property.
Numerous inscriptions show benefactions by individuals to communities. See VII. 24 for Ummidia Quadratilla's building of an amphitheatre and temple for the people of Casinum.

17 **sequenti:** *sc.* 'die'.

19 **continget hilares:** a difficult ellipse. We suggest it is short for 'continget
vos hilares invenire'.

Exploration

First, the journey. How long would a journey from Rome to Comum
take in Pliny's day? A journey of similar length, to Brundisium, took
Horace and his friends twelve days, though he admits they took it
slowly (*Satires*, I. 5). Today such a journey takes perhaps seven hours
by train, or less by fast car along the *autostrada*. Another comparison.
Nowadays we pick up the telephone and say we are on our way. Pliny
writes a letter. He and Calpurnia are going to travel as fast as they can
(lines 5-6). Is there any chance that they, as VIPs, may arrive before
the messenger carrying the letter?

Secondly, patronage. This letter shows (like IX. 39) that Pliny spent
money on building works. This was part of a patron's responsibility.
How much would a temple be likely to cost? (See X. 37; X. 39*, for
Pliny's mentions of particular sums of money being spent on public
buildings in Bithynia.) What did Pliny get out of building a temple at
his own expense? Modern parallels might be instructive, such as the way
firms and individual industrialists make handsome gifts to universities.

Thirdly, was this letter revised? The letter, as written, may merely
have been to say that Pliny and Calpurnia were looking forward to
visiting Fabatus, were just about to set off, but would have to turn
aside for a short time for the dedication of the temple at Tifernum-on-
Tiber. Revision for publication seems, however, to have given the letter
something more. It is inconceivable that Fabatus would need to be told
that Tifernum-on-Tiber was near Pliny's estates, in view of the way the
two men helped each other in their business interests (see above); and
Fabatus must also have known that Pliny was the patron of Tifernum
and that he was paying for the construction of the temple. (The reader
of Pliny's Letters is more than once told about this temple: see above
note on line 14.)

In its present, revised form this letter apparently intends to show
the readers (that is, the world) that Pliny puts his public duties as a
patron before his family commitments. However, in VII. 16 Pliny asks

Calestrius Tiro to turn aside from his journey out to Baetica as its new
governor, in order to manumit some of Fabatus' slaves — private business before public duty. There is even some similarity of wording:
'deflectemus in Tuscos' (IV. 1); 'ex itinere deflectat' (VII. 16). Does
Pliny mean these two letters to complement each other?

Questions

1 What relations is Pliny writing to?
2 How soon does he say that he and his wife are going to set off?
3 What place are they going to visit on the way? What special
   connection has Pliny with the place? How long has he had the
   connection?
4 What are they going to do there? Why is it an important duty?
5 How long will they stay there?
6 What tenses are 'contingat' (line 18) and 'continget' (line 19)?
7 Why does Pliny not mention what god the temple is for? Does
   he not think it important?
8 What advantages might there be for a town to have a patron in
   Rome?
9 Why should Pliny explain in such detail to Fabatus about his
   commitment at Tifernum?
10 Is this letter intended to pacify an irascible old man?

IV. 2

The letter

Pliny deplores Regulus' extravagant mourning for his son whom he
treated so despicably when he was alive.

It is not possible to identify Attius Clemens, to whom Pliny wrote
this letter, I. 10* and perhaps IX. 35* (S-W, 516).

This is one of several pen-portraits of Pliny's contemporaries. Often
such letters take the form of obituary notices (see, for instance, V. 5);
this letter however gives a sketch of a man still alive, M. Aquilius
Regulus.

Notes

1-2 **Regulus filium . . . putet:** this first sentence is saying three things:
    (1) 'Regulus filium amisit'; (2) 'hoc uno malo est indignus'; (3) 'sed
    nescio an malum putet'. With this division established, translation
    should not be difficult.

3 **posset ... referret:** one would expect these to be pluperfect subjunctives, and to make sense in English they must be translated as if they were so. The probable explanation is that 'posset' is not a conditional but a generic subjunctive, and 'referret' illogically follows suit in the same tense.

3-4 **hunc Regulus emancipavit:** a boy could not inherit while he was still under *patria potestas*. So Regulus had him legally freed from his control, so that he could wheedle his legacy from him. The legal process by which *patria potestas* was terminated was for the child to be sold three times to a third party who each time manumitted him.

5 **ex moribus hominis:** 'from knowledge of the man's character'.

6 **captabat:** see II. 20* for Regulus' insistent practice of legacy-hunting. It was common practice in the Roman life of the times for people to pay court to rich and childless people in the hope of inheriting something from them. See VII. 24, where certain people started currying favour with Ummidia Quadratilla. Pliny abominated Regulus' habit of legacy-hunting, and here suggests that he has reached the ultimate in baseness by paying court to his own son.

7-8 **amissum tamen luget insane:** Pliny continues the theme of Regulus mourning his son in IV. 7*.

10 **circa rogum trucidavit:** a conscious imitation of the funeral rites of heroes? Cattle were sacrificed round Patroclus' funeral-pyre (Homer, *Iliad*, XXIII. 165ff.) and also round the funeral-pyre in Virgil, *Aeneid*, XI. 197.

11 **convenitur:** i.e. for condolence.

16 **statuis suis:** people did not normally collect statues of themselves. Perhaps Regulus had an obsession with statues: he commissioned numerous statues of his dead son (IV. 7*).

18-19 **insaluberrimo tempore:** probably August, the most oppressive month in Italy.

19 **quod:** 'the fact that ... '

20 **hoc quoque sicut alia perverse:** 'and on this point too he is behaving as perversely as he does in everything else'.

23 **quo mendacius nihil est:** this use of 'nihil' when referring to a person is found elsewhere in Pliny, e.g. II. 9, lines 16-17: 'quo nihil verius nihil simplicius nihil candidius nihil fidelius novi'. IV. 22 has a more straightforward example: 'dixit Iunius Mauricus, quo viro nihil firmius, nihil verius' (lines 9-10).

Exploration

'The general colouring of the letters is amiable, sympathetic, improving. Censure of the living is the mildest imaginable' (Syme, *Tacitus* I. 96).

Pliny's hostility to Regulus is an exception to this generally benevolent treatment of people (see also I. 5*; I. 20*; II. 11*; II. 20* (sections 1-8 in *CLC* Unit IV 'domi'); VI. 2*, and *CLC Handbook* for Unit IV, p. 131). Regulus was a wealthy and successful barrister. He had prospered under Nero and, surviving the change of dynasty, became a prominent informer in the later years of Domitian's reign. Again he survived, dying during the reign of Trajan (S-W, 93-4; Syme, *Tacitus*, I. 100-2).

It is interesting to speculate on the reasons for this dislike. Political rivalry could hardly have been a motive, Pliny being much younger than Regulus and already an ex-consul: Regulus probably reached only the praetorship. There was obviously forensic rivalry, Pliny often crossing swords with Regulus in court, but this cannot be the root of the hostility: Pliny does not write shrill, abusive letters about his other legal rivals. It seems rather that the feeling is at a purely personal level. Pliny has conceived for Regulus a vivid personal hatred: he carefully keeps his distance from him but in fascination cannot resist being curious about everything that he does. It is an obsessive emotion which at times comes close to friendship: 'I often find myself missing Marcus Regulus in court, though I don't mean I want him back again' (VI. 2*). Even the present letter contains pathos as well as abuse. If Pliny had merely despised Regulus, would he have written such a letter at all, let alone gone into such detail?

Students may be interested to read, as a contrast, the epigram Martial wrote to Regulus' son when he was three and no doubt the hope of his parents:

> aspicis ut parvus nec adhuc trieteride plena
> Regulus auditum laudet et ipse patrem?
> maternosque sinus viso genitore relinquat
> et patrias laudes sentiat esse suas?
> iam clamor centumque viri densumque corona
> volgus et infanti Iulia tecta placent.
> acris equi suboles magno sic pulvere gaudet,
> sic vitulus molli proelia fronte cupit.
> di, servate, precor, matri sua vota patrique,
> audiat ut natum Regulus, illa duos.
> (VI. 38)

Iulia tecta (line 6): the Basilica Julia was the place of the Centumviral Court.

Questions

1 What legal trick had Regulus used with his son? For what purpose?
2 How was the funeral unusual?
3 Were the condolences which he was receiving genuine? What was the motive for them?
4 What contradictions does Pliny allege in Regulus' character?
5 'vexat . . . civitatem' (line 18): do you think this is a sensible comment?
6 Explain what is meant by 'quorum alterum immaturum alterum serum est' (lines 21-2).
7 How does Pliny summarise Regulus' character?

8 Do any parts of the letter contain pathos? What impression do
   you have of Regulus, from Pliny's description?
9 Why do you think Pliny makes this the subject of a letter?

IV . 16

The letter

Pliny has made a successful speech before the Centumviral Court, and is
pleased that recognition is still given to good oratory.

Pliny wrote five letters to Valerius Paulinus, of which two (this and
V. 19) are included in this Selection. Valerius Paulinus had estates in
Gallia Narbonensis at Forum Iulii (V. 19), so he was presumably the
son of that citizen of Forum Iulii and procurator of Narbonensis who
had given the region's support to Vespasian in AD 69 (Tacitus *Histories*,
III. 43). Probably because of this, Valerius Paulinus was able to pursue
a successful senatorial career, being consul in AD 107. He died *circa*
AD 110-11 (X. 104*).

Notes

2 **studiis:** Pliny here means specifically 'oratory'.
2-3 **apud centumviros:** for the Centumviral Court, see V. 9*; VI. 33*,
   and *CLC* Handbook to Unit IV pp. 144-5, as well as general works of
   reference (e.g. *OCD*). The court dealt with civil lawsuits, in particular
   cases of inheritance. It originated in Republican times and was placed
   by Augustus under the general supervision of the *praetor ad hastam*
   (or *hastarius*). The court was by Pliny's time composed of 180 judges;
   usually they did not all preside together over a case, but see IV. 24*
   and VI. 33*. The Centumviral Court could help a man make his name;
   and Pliny seemed to think it had played an important part in his
   career (IV. 24*).
5 **tunicis:** this word is used indiscriminately in the singular or plural.
7 **horis septem:** Pliny had spoken for just about a whole day of the
   court's time, the normal length of time for a speech apparently being
   about an hour.
   tam diu: these words refer back to 'horis septem': 'for this was how
   long I spoke . . . '
7-8 **maiore cum fructu:** this kind of hint is the nearest Pliny gets to discussing the tangible results of his court cases.
9 **sunt qui audiant . . . legant:** generic subjunctives: 'there are people
   ready to listen and people ready to read'.

Exploration

The interest of this letter centres around Pliny's reasons for including
it among those he published. He was proud of his career in the Centumviral Court, but the emphasis here is on court speeches as a branch of

literary studies. In I. 13 he deplores the uncouth behaviour of audiences at *recitationes*: the listeners are not interested, whereas here a young man stays for seven hours in discomfort to hear a speech.

Other questions arise, but discussion of them can only be speculative: for instance, we cannot know whether the court was crowded because Pliny was a popular lawyer or because this was a *cause célèbre* or because the court always was crowded (cf. VI. 33*); nor can we tell whether the young man stayed to hear Pliny or for some other reason.

General discussion could include the difference between civil and criminal courts, and the place of oratory in Roman life.

Questions

1 What happened to make Pliny conclude that 'adhuc honor studiis durat'?
2 How long did Pliny speak?
3 Why was it particularly uncomfortable for the young man to be dressed only in a toga? What was a toga made of?
4 Did the young man stay necessarily for the love of oratory?
5 Does Pliny imply that he won the case? In what words?
6 What does he conclude from the incident? Does it apply just to public speeches?
7 Given that the Romans were inclined to boasting, is Pliny rather too pleased with himself?

IV. 22

The letter

Pliny recounts a meeting of Trajan's *consilium*, which reminds him of an earlier dinner-party with Nerva, because Junius Mauricus showed courage on both occasions.

Sempronius Rufus was *consul suffectus* in AD 113; but nothing further is known of his career.

Notes

1 **principis optimi:** i.e. the Emperor Trajan.
consilium: for the emperor's *consilium*, see *CLC* Handbook for Unit III, pp. 134-5 and particularly J. Crook, *Consilium Principis* (Cambridge U.P., 1953). For Pliny's attendance at the *consilium principis,* see also VI. 22* and VI. 31*. The *consilium principis* did not have a fixed number of permanent members, but suitable people (like Pliny) might be called.

2 **gymnicus agon:** Greek words, because this was a festival in the Greek style, with athletics, dancing and music: an example of the influence of Greek culture in Gallia Narbonensis. Agricola had been educated in Massilia (Marseilles) and Tacitus praises the city as 'locum Graeca comitate et provinciali parsimonia mixtum ac bene compositum' (*Agricola,* 4).

apud Viennenses: Vienna (modern Vienne) was in Gallia Narbonensis. It was capital of the Allobroges. Gaius Caligula gave it the status of a Roman colony, its magistrates being after this grant *duoviri* (S-W, 299).

4 **tollendum abolendumque:** not synonyms, because the first word implies immediate, the second prospective, action: 'took steps to have this cancelled and abolished'.

5 **negabatur ex auctoritate publica fecisse:** it is a little surprising that a matter of such comparative unimportance in a town in a public province should have found its way to the emperor's *consilium,* though by this time the emperors were concerning themselves more with public as well as imperial provinces (see Fergus Millar, 'The Emperor, the senate and the provinces'). However, this dispute was bound to find its way to the emperor, whether he had been directly appealed to or whether it had come through the channels of the proconsul of the province and then the senate. The point at issue called into question something which applied to the entire empire: did a *duumvir* have the power to annul the will of a Roman citizen? Pliny was present as one of the most trained legal minds of his day.

7 **tamquam homo Romanus:** though not a Roman by birth, Trebonius behaved like one. The people of Vienne were Gauls who had Roman citizenship (see note on line 2).

7-8 **in negotio suo:** the sense is 'on a matter which touched him personally'.

8-9 **sententiae perrogarentur:** the emperor asked everyone's opinion but did not bind himself to accept the view of the majority.

9 **Iunius Mauricus:** he was in the senate by the end of Nero's reign. He was exiled by Domitian in AD 93 but returned in AD 97. Pliny knew him quite well: he arranged for the betrothal of his niece (I. 14*); he found a tutor for his nephews (II. 18*); and he stayed with him at Formiae (VI. 14*).

nihil firmius: for this use of 'nihil' in reference to a person, see II. 9.

10 **agona:** Greek Form B (accusative) singular.

11 **vellem etiam Romae tolli posset:** Domitian had instituted at Rome a festival on Greek lines called the *Quinquatria* (S-W, 299). It was to this that Mauricus was referring: he was certainly not suggesting that the Roman *ludi* or *munera* be abolished. Some Romans objected to Greek games on the grounds that men appeared naked in public and they were associated with homosexuality (S-W, 301).

constanter: *sc.* 'loquebatur'; so also with 'fortiter' (line 13).

14-15 **Veiento proximus . . . recumbebat:** Veiento was reclining on the Emperor's right, the place for the guest of honour. (For the seating arrangements at a Roman dinner, see Paoli 92-3.) But why is Veiento's presence important? Because he had been one of Domitian's counsellors and had retained his favoured position under Nerva. His presence as the latter's honoured guest made Mauricus' frankness even more courageous; or perhaps it was his presence which induced the remark anyway.

Like quite a number of his senatorial contemporaries, Fabricius Veiento had a chequered career. Nero banished him in AD 62 for his satirical *Codicilli.* Returned from exile, he was influential throughout

the Flavian period. Opinions differ about whether he was a prominent
delator (informer) during Nero's reign. Certainly this letter shows that
he had Nerva's confidence. He was consul three times.

16 **de Catullo Messalino:** L. Valerius Catullus Messalinus was consul in
   AD 73 and shared his second consulship (AD 85) with the Emperor
   Domitian. He obviously enjoyed Domitian's trust. Tacitus names him
   as one of Domitian's *delatores* (*Agricola,* 45). This present letter
   suggests that his influence in Domitian's *consilium principis* was considerable. Juvenal has him attending the emperor's *consilium* on the
   matter of the giant turbot (see 'Exploration'), and gives as unfavourable
   a portrait of him as Pliny does here, for instance, in these three lines:

   et cum mortifero prudens Veiento Catullo,
   qui nunquam visae flagrabat amore puellae,
   grande et conspicuum nostro quoque tempore monstrum.
                                           (*Satires,* IV. 113-15).

18-20 **quo saepius . . . contorquebatur:** to begin with, students should estab-
      lish that 'non secus ac . . . feruntur' is the simile, and this should be
      left until the rest of the sentence is dealt with.
   19 **quae et ipsa . . . feruntur:** 'for they too fly to their mark blindly and
      indiscriminately'.
   21 **sententiis:** this word refers to opinions in the Emperor's *consilium*
      (line 8) and also opinions given in the senate (S-W, 301).
 22-3 **passurum fuisse:** an almost colloquial use of tenses: 'What do we think
      might have happened to him if he were still alive?'
   23 **nobiscum cenaret:** a wickedly barbed remark to make in the presence
      of Veiento (see note on lines 14-15).
 23-4 **longius abii:** this is Pliny's acknowledgement that he has broken his
      own custom of dealing with only one subject in each letter.
   24 **placuit:** an official term, often used of senatorial resolutions: here, of
      the resolution of the Emperor's *consilium.* Translate 'it was resolved...'
 24-8 **placuit agona tolli . . . :** in particular the words 'ut noster hic omnium'
      (line 25) are rather elliptical. A translation of the final four lines of
      this letter may help with the difficult Latin. 'It was resolved to stop
      the games which had corrupted the morals of the people of Vienne
      just as our games in Rome corrupt the morals of the world. For while
      the vices of the people of Vienne remain among those people them-
      selves, our vices travel far and wide, and just as in human bodies, so in
      an empire the most serious disease is that which spreads from the
      nerve-centre.'

Exploration

This letter has no single theme but it describes the intimate circles of
two emperors and Junius Mauricus is present on both occasions.

'The rule of the Caesars . . . announces the era of cabinet government' (Syme, *Tacitus,* I. 5). Cyrene Edict V (EJ 311) shows that even
as early as 4 BC the real decisions were taken by a *consilium*, in that
case the *consilium semestre* as against the *consilium principis* here. The
real influence was behind the closed doors of the palace and it was
proximity to the emperor that really counted. It had been so in
Claudius' reign: the freedmen had been influential not because of their
office but because they had the ear of the Emperor. So, in its way, an

imperial dinner-party was nearly as influential as an imperial *consilium.*

This private cabinet of the emperor's is satirised at length by Juvenal in the well-known tale of the turbot (*Satires* IV), in which the various counsellors are summoned to Domitian's villa at Alba Longa, no doubt expecting to discuss some momentous matter of state, only to find that the emperor wants a decision on the fate of a giant fish with which he has been presented. Juvenal presents the members of the *consilium* as being men in a powerful but perpetually dangerous position: they are

> proceres, quos oderat ille,
> in quorum facie miserae magnaeque sedebat
> pallor amicitiae.   (73-5)

Both Catullus Messalinus and Veiento appear in the Satire. What did these men gain from being in such a position and having to act as advisers, informers, murderers or perhaps eventually scapegoats for an emperor who seems to have felt little more than contempt for them? The answer seems to have been that the power was something for which these men were willing to run risks, and some of them, like Veiento (whom Juvenal calls 'prudens'), managed to survive by never saying or doing the wrong thing.

However, to be in Trajan's *consilium* was a different thing, to judge from Pliny, who seems to have no fear about its dangers and even publishes a letter recounting a recent meeting of it: and the frank remarks of Mauricus in the presence of both Nerva and Trajan seem to indicate that the *consilium* is now a genuine meeting of advisers, and both here and at dinner-parties it is safe to say what one feels.

Pliny's hatred of Domitian's régime is clear throughout this letter. The very publication of such an account is an implicit criticism of Domitian's secrecy and intrigues; Veiento and Messalinus are condemned, the one by implication, the other in violent language; and finally Pliny says that Domitian's Quinquatrian games have corrupted morals everywhere (lines 24-8). Furthermore, many of Pliny's readers would remember that Mauricus was the brother of Arulenus Rusticus (for whom, S-W, 740), murdered by Domitian.

Students may find it interesting to discuss the disapproval that existed of the Greek style of games although they were so much more civilised than the sordid and even sadistic *munera* to which many of the population were addicted (and on which even Pliny could bestow praise — VI. 34). Was this 'straining at a gnat and swallowing a camel'?

Questions

1 Who is 'principis optimi' (line 1)?
2 What kind of games do the words 'gymnicus agon' suggest?
3 Why were games of that kind popular at Vienne?

4 Why was Trebonius' action a serious matter? Who do you think had objected to it?

5 Was Trebonius a Roman by birth?

6 What was courageous about the remark 'vellem etiam Romae tolli posset'?

7 Explain the significance of 'Veiento proximus atque etiam in sinu recumbebat' (lines 14-15).

8 Why does Pliny mention Veiento?

9 What had made the character of Catullus Messalinus even more unpleasant?

10 How did Domitian use him? What simile does Pliny use to describe this? Do you think it is appropriate?

11 Why was 'nobiscum cenaret' such an outspoken remark? Which of those present was it likely to offend?

12 In what way was a 'gymnicus agon' likely to corrupt people's morals? Was it more likely to do so than Roman-style games? Or did it do so in a different way?

13 What word in the last sentence does Pliny use with a double meaning?

14 What custom does he break in this letter? Where does he admit that he has done so?

15 How many subjects does this letter deal with?

16 Is Pliny being daring to publish this letter? Does it present Trajan's *consilium* in a favourable light? Does it present Nerva in a favourable light?

17 Are there many circumstances in which it is right to annul someone's will?

18 Would you describe Greek-style games as more civilised than Roman? Is Pliny simply condemning them out of prejudice?

19 Were Romans accustomed to appearing naked in public? In what public buildings did they do so?

V . 5

The letter

Pliny writes feelingly about the death of Gaius Fannius, who as a result of seeing Nero in a dream had an accurate premonition that he would not finish his great work on Nero's political victims.

Novius Maximus, to whom Pliny also wrote IV. 20* and perhaps other letters, cannot be identified for certain (S-W, 297).

Other 'pen-portrait' letters in this Selection: VI. 16 (which describes *inter alia* the death of the Elder Pliny) and VII. 24 (Ummidia Quadratilla).

Notes

1   **C. Fannium:** Gaius Fannius cannot be identified. He was perhaps related to Fannia, the daughter of Thrasea Paetus and the wife of the elder Helvidius Priscus (Syme, *Tacitus*, I. 92). The family had a tradition of opposition to the emperors, suffering under Nero (see, for instance, B. H. Warmington, *Nero, reality and legend,* Chatto & Windus, 1969, 142ff.); so the subject-matter of Fannius' book may have been especially appropriate.

5   **veteri testamento:** Form E (ablative) absolute, if such classification is possible here; 'veteri' is less commonly used than 'vetere'. Cf. VII. 24 of Ummidia Quadratilla: 'decessit honestissimo testamento' (line 4).

It is impossible to say when Fannius had last made a will but the exclusion of 'quos maxime diligebat' (line 6) from it suggests that it was some time back. In VIII. 18* Pliny writes about Domitius Afer, who left a will that had been drawn up eighteen years previously. Romans may on the whole have changed their wills regularly, and this gave the legacy-hunters their chance (S-W, 320). See J. A. Crook, *Law and life of Rome* 118ff for Roman wills and the question of inheritance.

9-10 **exitus occisorum aut relegatorum a Nerone:** many prominent people fell victim to Nero: his own mother, Agrippina the Younger (Tacitus, *Annals,* XIV. 8, in *CLC* Unit V); Claudius' son Britannicus, whose death was generally accepted to be by poisoning (XIII. 16); nineteen people killed and thirteen exiled as a result of the Pisonian conspiracy of AD 65 and its aftermath, among these Seneca (XV. 60-4), and his nephew Lucan (XV. 70). Petronius 'Arbiter' committed suicide (XVI. 18-19), as did Thrasea Paetus (XVI. 34-5). The elder Helvidius Priscus was exiled (XVI. 33), being banished again by Vespasian and subsequently executed (Suetonius, *Divus Vespasianus,* 15). Corbulo, summoned to Greece by Nero, killed himself on the Emperor's orders (Dio Cassius, LXII. 17).

10-12 **tres libros absolverat subtiles . . . medios:** Pliny, with his great interest in literature and his own experience of writing, feels able to comment on Fannius as a writer.

11 **Latinos:** implies that Fannius, like Pliny, strove to preserve the classical purity of style and diction.

18 **nulla mors non repentina est:** the double negative (so dear to the Romans) needs cancelling out to become 'always'.

18-19 **ut quae . . . abrumpat:** 'because it always breaks off something that

has been started'. 'nulla mors' is the antecedent of 'quae'. Once this has
been established, 'ut quae' can perhaps be translated: 'as something
which . . . '

21-2 **iacere . . . scrinium:** evidently Fannius was writing his own manuscript.
Pliny also lay down when he worked at his writings, which he dictated
to a secretary (IX. 36*).

26-7 **expavit et sic . . . legendi:** 'he was terrified and interpreted it as if he
would have the same stopping-point in the writing as Nero had had in
the reading'.

28-9 **quod me recordantem . . . :** 'when I think about this I am filled with
pity for how many sleepless nights and how much effort he used up in
vain'.

32-3 **enitamur ut mors . . . :** 'let us strive that death may find as few things
as possible to destroy'.

Exploration

Some twelve of Pliny's letters may be considered as obituary notices.
The very first sentence here, 'nuntiatum mihi C. Fannium decessisse',
announces that this is one such. Pliny is much grieved at Fannius' death:
he held him in affection and respected his good taste, eloquence, sound
judgement and native intelligence (lines 1-14). Pliny and he may have
known each other well, as Fannius was a lawyer ('quamvis enim agendis
causis distringeretur': lines 8-9) and a writer. It is a quiet reminder of
the gaps in our knowledge of the ancient Roman world that we cannot
identify Fannius, beyond what Pliny tells us here and the possibility
too that he was connected with the family of Thrasea Paetus (see note
on line 1).

It is easy to see why Pliny wrote some of his obituary letters. II. 1*
was obviously written as an act of *pietas:* Verginius Rufus, to whom it
pays tribute, had not only been a distinguished public figure (something Pliny acknowledges) but also Pliny's guardian ('he gave me a
father's affection'). VIII. 23* mourns the untimely death of Junius
Avitus, whom Pliny had helped so much in the way he himself had been
helped by others like Verginius Rufus. But what is Pliny's purpose in
writing about Fannius? The letter itself may provide two answers:
Firstly, Fannius' will, being an old one, omitted to mention some of his
dearest friends. This was a cause of distress to Pliny (lines 4-6), who
remarks elsewhere that 'there is no truth in the popular belief that a
man's will is a mirror of his character' (VIII. 18*) and who may now
be wanting to set the record straight about Fannius. Secondly, there is
this notion of Pliny's that Fannius will not win immortality through his
writings, because they were not finished. Fannius was one of those to
whom death had come prematurely when they were planning some
immortal work (lines 14-15), and Pliny uses the word 'frustra' (line 29)
of all the hours Fannius had spent on his writings. It is no use our
quarrelling with Pliny's view (will Fannius not have gained immortality
even from his unfinished work? one could ask): it was Pliny's belief that

it had all been in vain; and one suspects that he wanted through this letter to give Fannius the immortality of which he had been cheated by death. Such harmless (and not unfounded) conceit on Pliny's part is balanced by his generosity of intention.

The little ghost story in this letter is not the only one in Pliny: three occur in VII. 27* (the spirit of Africa, the haunted house at Athens (simplified version in *CLC* Unit III, Stage 31) and the story of the hair being cut at night (in *CLC* 'Information about the language IV/V', p. 50). Pliny draws no connection between ghosts and the possibility of life after death. For him, it seems, immortality rests in being remembered (as this letter suggests), while other writers such as Catullus and Horace sadly contrast the birth of a new day or a new month or year with the one life each human has. So Catullus writes:

> soles occidere et redire possunt:
> nobis cum semel occidit brevis lux
> nox est perpetua una dormienda.   (V)

Or there is Horace, *Odes* IV. 7: students could read this beautiful poem either in Latin or in one of several English translations (in particular, that of A. E. Housman, 'The snows are fled away . . . ').

Questions

1 What were the two reasons why Fannius' death was particularly distressing?
2 What was his profession?
3 What kind of work was he compiling?
4 What words tell us that some of it had already been published?
5 Was the part which had been published popular?
6 Does Nero seem to have killed or banished many people?
7 For what kind of people does Pliny say death is not such a great disaster?
8 What gave Fannius a premonition of his death? Does Pliny take it seriously?
9 How does the story of the premonition seem to contradict something earlier in the letter (see line 5)?
10 Why does Pliny think that Fannius' work and sleepless nights were in vain?
11 What reflection does the incident induce in Pliny himself?
12 Which sentence makes it clear that the recipient of the letter is also a writer?
13 Do you agree that an unfinished work of art is necessarily worthless? Do you know of any other unfinished works of art?

14 What does Pliny seem to think is the only way of gaining im-
    mortality?
15 Do you find this a depressing letter?
16 Give an example from one of the other letters of Roman
    belief in omens (see III. 14).

V.19

The letter

One of Pliny's freedmen is chronically ill, and so Pliny is sending him
for recuperation to Paulinus' estate in Provence.

The other four letters in this Selection concerned with Pliny's
humanitas are II. 6, VII. 16, VIII. 16 and X. 31.

For Valerius Paulinus, see IV. 16.

Notes

2 **qua indulgentia:** it may help to point out at the beginning that these
   words are Form E (ablative).

3 πατὴρ δ'ὡς ἤπιος ἦεν: this is the remark made by Telemachus about his
   absent father Odysseus (*Odyssey*, II. 47), remembering how he treated
   his subjects.

4 **pater familiae:** this is the later form of the archaic expression 'pater-
   familias' which uses the old genitive ending in -s. The expression, as
   Pliny makes clear, represented a whole ideal of fatherliness which em-
   braced sternness, tenderness and nobility; cf. Seneca, talking of the past:
   'dominum patrem familiae appellaverunt, servos . . . familiares' (*Letters*
   XLVII. 14).

5 **infirmitas liberti mei Zosimi:** Zosimus was evidently suffering from
   consumption. Compare the illness of Pliny's *lector* Encolpius
   (VIII. 1*).

5-6 **cui . . . eget:** 'illa' refers to 'humanitas' and is Form E (ablative) after
   'eget'. The sense is 'to whom so much more kindness must be shown
   now that he needs kindness so much the more'.

7-8 **et ars quidem eius . . . comoedus:** only an apparent syntactical illogi-
   cality. Pliny is quoting the word 'comoedus' as if he is reading Zosimus'
   label. 'In fact his profession, his label, one might say, is "actor".'

8 **comoedus:** such a slave or freedman would be a trained actor, used to
   performing in a troupe. But Zosimus went beyond his training and
   was accomplished at reading and playing the lyre (the training of a
   lector and a *citharoedus*). His duties in Pliny's household might be tc
   give readings and musical interludes after dinner; he might also read to
   Pliny personally, a frequent duty of such slaves.
   in qua plurimum facit: 'in which capacity he is extremely accomplished'.

19 **in Aegyptum missus:** probably Zosimus went to Egypt, when Pliny's
   friend Vibius Maximus (to whom he wrote III. 2*) was *praefectus*

Aegypti (AD 103-7). Sherwin-White notes that Celsus and the Elder
Pliny recommended Egypt for the healthy voyage there (S-W, 351:
210 for Vibius Maximus).
23 **Foro Iulii:** the modern Fréjus, in Provence.
26 **villa:** *sc.* 'pateat'.
    offerant: *sc.* 'tui'.
29 **frugalitate:** Form E (ablative) of cause: 'because of his temperance'.
29-30 **proficiscenti . . . eunti:** *sc.* 'ei' or 'Zosimo'.

Exploration

The opening sentence, and indeed the request to Paulinus, indicates that
the two men shared a humane attitude to slaves and freedmen. In his
will Paulinus named Pliny as patron of his Latin freedmen (X. 104*),
perhaps reckoning that they would receive better treatment from Pliny
than from his own son. Pliny and Paulinus were liberal for their day;
but their attitude to slaves and freedmen had been foreshadowed a
generation earlier by Seneca (see, in particular, *Letters* XLVII).
    It would be useful to know more about Pliny's association with
Paulinus. When we consider that Paulinus was much more than a provincial nobody — he had a successful political career and was *consul
suffectus* in AD 107 — Pliny seems a shade high-handed in his plans for
Zosimus' stay on Paulinus' estates: 'destinavi' (line 22) does not seem
to allow Paulinus to say 'No' to Pliny — compare the use of 'destino' in
III. 6 (line 23); and then Paulinus' household is to meet Zosimus' expenses. But a Roman may have found Pliny's tone much more in order.
For instance, 46 BC finds Cicero writing to the governor of Greece in
not dissimilar terms about a freedman of one of his friends: 'You will
therefore be doing me a favour, if you admit him into your friendship
and help him if he needs it in any way, without any inconvenience to
yourself' (*Ad Familiares* XIII. 23).
    In making such calls upon hospitality, neither Cicero nor Pliny were
acting unconventionally: it was implicit in such requests that the writers
would reciprocate when asked.

Questions

1 What word suggests that Pliny is slightly embarrassed about
   the way he treats his slaves? Or is he affecting embarrassment?
2 Of what nationality do you think Zosimus is?
3 What is Zosimus' job in Pliny's household?
4 What words does Pliny use to express how much he values
   Zosimus' services?
5 What general maxim does Pliny state which fits the situation?
6 What is Zosimus suffering from?
7 What famous Roman soldier was born at Forum Iulii?

8 Which part of the letter suggests that Pliny is on very intimate terms with Valerius Paulinus?

9 How do you know that Zosimus repays Pliny's concern for him?

10 Did freedmen always remain with their former masters?

VI. 7, VII. 5

The letters

Pliny writes to his wife Calpurnia to say how much he is missing her.

Notes: VI. 7

2 **libellos meos**: IV. 19* discloses that Calpurnia kept copies of Pliny's works to read again and again and learn by heart.

3 **in vestigio meo**: this may mean in the marriage-bed, but it is not clear.

6-8 **nam cuius litterae . . . inest**: this is more easily translated if the order of the two clauses is reversed.

9 **delectet . . . torqueat**: these are subjunctives after 'licet', and have nothing to do with 'ut', which takes up the earlier 'ita': 'though this tortures me as well as delights me'.

Notes: VII. 5

5 **diaetam**: a Greek word meaning a suite or apartment, sometimes even self-contained, which was used in the daytime.

6 **similis excluso**: 'like a man shut out'.

9 **in miseria curisque**: i.e. of Pliny's day-to-day legal and public duties.

Exploration

Pliny published three letters written to Calpurnia (the present two and VI. 4*). We read in VI. 4* that Calpurnia had gone to Campania for her health, and a reluctant Pliny, detained by public duties, was unable to join her there. This was the letter of an anxious husband, anxious because any man in love fears the worst when he has not heard from his wife, and in Pliny's case anxious also because Calpurnia is in poor health. So he bids her write 'once or even twice a day'.

By the time Pliny comes to write VI. 7, Calpurnia has written back, and more than once, it seems (lines 4-5). Showing her acceptance of the point that Pliny is busy with an official round of duties, she has spoken of being comforted by holding his *libelli* to her (these are probably his speeches, not — one presumes — his letters, nor petitions like those which unwelcome country-people push at Pliny during a stay in Tuscany: IX. 15). She knows her husband: if he cannot be with her, at least he will be delighted that she has his writings close to her.

VII. 5 follows easily on from the other two letters. In VI. 4* it was Pliny who complained of his absence from Calpurnia; in VI. 7 the theme was how much Calpurnia misses him: 'scribis te absentia mea non mediocriter adfici' (line 1); and in the third letter Pliny goes back, more briefly but perhaps more intensely, to how much he misses Calpurnia. Always an artist, he ends VII. 5 with the words 'cui requies in labore, in miseria curisque solacium'. It was his work that was keeping him from Calpurnia we were told originally (VI. 4*), but it is in his work that he finds, in the final letter, consolation for his separation from her.

If the artistry is evident in the structure of the letters, it may also be present in the content. 'excluso' (VII. 5, line 6) may be taken as evidence of the idea of Pliny as an *amator exclusus* (for reference to which, see S-W, 359). The *amator exclusus* is a convention of Latin literature, and Pliny is allowing himself to be whimsical here: he is *exclusus* and by the tone of his letter an *amator;* but he is *exclusus* from an empty room — Calpurnia being away in Campania.

However far one goes along with the idea of artistry in these three letters, it must not be allowed to prejudice our judgement of their sincerity. The very fact that Pliny writes so openly about his commitments in Rome (commitments which he once told Caninius Rufus had the effect of being like chains: II. 8) proclaims the honesty and hence the sincerity of these letters. Like other Romans of his background, Pliny had been trained to accept the demands of duty; and it could only be a man who recognised the overriding claims of these demands and did not question them, who would dare to write to a wife he loved and tell her that this was why he could not be with her. But, if he acted in this way, he recognised that Calpurnia must do so too: when her grandfather died, he let her leave him in Bithynia and go to comfort her aunt in her bereavement, even though this duty would force on them a separation such as the three letters under discussion here had earlier forced on them (X. 120*).

In these three letters we come very close to Pliny, to a man who had such a highly developed sense of duty, who loved his wife so unaffectedly and who is perhaps revealed here as a less urbane, more sensitive and more emotional man than many of his polished letters would have us believe.

Questions

1 How does Calpurnia console herself for the absence of her husband?
2 Is it necessary for them to be separated?
3 Which words in VII. 5 imply that there was a fixed period of the day which Pliny and his wife normally spent together?

4 What is the only time when Pliny forgets that he is missing his wife?

5 Which word does Pliny use in both letters to express his emotion when his wife is not there?

VI. 16

The letter

In response to a request from Tacitus, Pliny gives an account of his uncle's death during the eruption of Vesuvius.

For a discussion on Pliny's friendship with Tacitus and details of other letters to him, see I. 6.

For other 'pen-portrait' letters, see V. 5.

Notes: sections 1-6 (English translation)

My uncle: Pliny the Elder was the brother of Pliny's mother. He adopted Pliny, thus making him his legal heir, and so the young man took the name Plinius and inherited his uncle's full estate.

Pliny the Elder was an *eques,* unlike his nephew who pursued a senatorial career. He was born in AD 23/4, served in various equestrian posts and was at the time of his death in charge of the fleet at Misenum. He had practised as a barrister and was also a writer: for a list of his writings and a description of his remarkable daily routine, see III. 5*.

the twenty-fourth of August: this gives us our knowledge of the exact date of the eruption. Suetonius (*Divus Titus,* 8) merely places it in Titus' principate; and Dio, epitome of LXVI, 21, puts it in Titus' first year as emperor.

had been sunbathing: in III. 5* Pliny records that his uncle would often in summer recline in the sun and have some author read to him.

I believe this was because . . . : Pliny arrives at the correct explanation — 'a worthy pupil of his uncle' (S-W, 373).

Notes: 7-20 (Latin text)

1 **magnum:** 'important'.

2 **liburnicam:** one of the fast vessels belonging to the Roman fleet.

3 **respondi studere me malle:** for Pliny's interest in intellectual pursuits, see I. 6.

4 **codicillos:** line 6 makes it clear that the message must have come by boat, in which case why had Rectina not escaped there and then? Presumably she wanted a large boat to come and take away some of her belongings.

5 **Rectinae Tasci:** Rectina is the wife of Tascius Pomponianus (line 19). Pliny does not give the location of her villa, though 'subiacebat' (line 6) suggests the region of Herculaneum. It becomes clear later in the letter

that the Elder Pliny gives up the attempt to reach her when he finds the boat can get no nearer (lines 16-19), and instead he goes to rescue her husband who is at Stabiae. Rectina's fate is unknown. It is interesting that Pliny mentions these people's names in a way which suggests Tacitus knows all about them.

8 **maximo:** 'heroic'.

deducit quadriremes: on receiving the message from Rectina, Pliny the Elder, who is in command of the fleet, decides to use warships to rescue as many people as possible. These are heavier ships than the *liburnica* (line 2) which he initially ordered for his personal use to make observations.

9-10 **amoenitas orae:** an abstract idea substituted for a more concrete phrase.

12 **solutus metu:** 'devoid of fear'.

14 **quo propius accederent:** 'the nearer they approached'.

16-17 **iam vadum subitum . . . :** 'now suddenly the water was shallow and the shore blocked their way with debris from the mountain'.

19 **Pomponianum:** the Elder Pliny had written a life of Pomponius Secundus (III. 5*) who had a son called Tascius Pomponianus — presumably the man mentioned here, who is the husband of Rectina (line 5).

19-21 **Stabiis erat . . . :** 'he was at Stabiae, cut off [i.e. from his wife] by the intervening gulf (for the sea washes into a gently enclosed, circling coast)'. The Elder Pliny had originally intended to land somewhere near Herculaneum, where Rectina apparently was; prevented from doing that, he now decided to pick up her husband, who happened to be at Stabiae on the inner side of the inner bay.

22 **proximo:** *sc.* 'futuro': 'likely to come very close'.

24 **quo:** connecting relative: 'to that place', i.e. Stabiae.
secundissimo: *sc.* 'vento',

27 **aeque magnum:** 'equally heroic'.

29 **latissimae . . . incendia:** 'great sheets of fire and soaring flames'.

30 **agrestium:** Strabo (V. 247) commented that Mount Vesuvius had dwellings all around it, except at the summit, and that the farmlands were superb. He remarks too that the fertility of the land may be connected with earlier volcanic activity.

33-4 **meatus animae . . . gravior et sonantior erat:** this and the subsequent circumstances of the Elder Pliny's death suggest that he suffered from asthma.

39 **in commune consultant:** 'together they debated whether . . . '.

43 **quamquam:** governs only the two adjectives: 'the fall of pumice stones was feared, even though they were light and porous; however, a comparison of the dangers suggested the latter choice'.

49 **placuit:** 'it was decided'.

51 **vastum:** often a difficult word to translate: perhaps 'swollen'.

52-3 **frigidam aquam poposcit:** apparently he was already beginning to choke.

58 **is ab eo . . . :** 'this was the second day after the last day he had seen'. Pliny's reckoning is inclusive, as is usual with Romans. The time-sequence is:

24 August:	The Elder Pliny leads rescuers as far as Stabiae. Daylight.
25 August:	No daylight. Death of the Elder Pliny and presumed escape of others.
26 August:	Discovery of the body.

59 **corpus inventum integrum inlaesum**: it is possible that Pliny is emphasising that his uncle was definitely not murdered by his slaves; but,
more likely, he simply wishes to establish how peacefully his uncle
died.

Exploration

This is one of the best known of all Pliny's letters. Tacitus had asked
Pliny (who had been only seventeen at the time of the events described:
see VI. 20*), to send him an account of his uncle Pliny the Elder's
death to incorporate in his own writings. This letter is an answer to that
request: and it is a graphic description of the eruption of Vesuvius that
destroyed Pompeii and Herculaneum and much else in the area. Another
letter, VI. 20*, was written in answer to Tacitus' invitation to Pliny to
describe his own adventures at the time of the eruption. Tacitus'
account of the eruption has not survived (we have less than five books
of his *Histories*) but he does remark in the first book: 'haustae aut
obrutae urbes, fecundissima Campaniae ora' (I. 2). Tacitus evidently
recognised Pliny's usefulness as a source: his descriptions are simple and
vivid. For another account of a natural disaster, see VIII. 17, on the
flooding of the rivers Anio and Tiber.

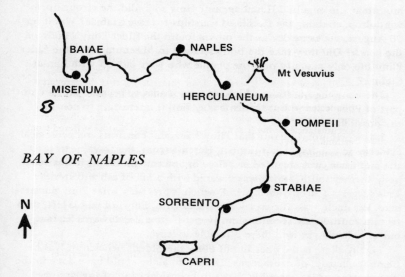

It is a challenging task to follow the story — partly because Pliny
assumes Tacitus' acquaintance with the topography and with the two
people mentioned, Rectina and Tascius, and partly because the modern
reader has to remind himself continually about the dependence of

ancient ships on a favourable wind. If we accept the theory (S-W, 373) that Pomponianus (line 19) is Tascius Pomponianus and therefore the husband of Rectina (line 5), then the course of events becomes intelligible. The Elder Pliny sets off from Misenum with his rescue-flotilla, sailing eastwards towards the foot of Vesuvius: he is intending to rescue Rectina and others in the region round Herculaneum. The wind, blowing from the north, is on his left. Suddenly the ships can go no further (lines 16-17), because the water has become too shallow (Pliny implies in VI. 20* that the sea-bed had risen). The helmsman suggests that they turn back. Pliny the Elder agrees that they cannot reach that part of the coast but decides to turn south to pick up Pomponianus at Stabiae: he will get there quickly with a following wind and is gambling on the wind changing in time for them to put out again from Stabiae before too long. This is the gamble which he personally loses, because during the wait at Stabiae for the wind to change, his weak chest causes him to die of suffocation.

The account leaves several questions unanswered. Why did Rectina not leave immediately? Someone was able to leave her house to take the message, so why did she not escape herself (lines 4-7)? What was the fate of Rectina, since the Elder Pliny could not reach her? Did she perish under the wave of volcanic mud which swept down from the mountain and engulfed Herculaneum? How soon did the strong north wind drop, enabling the flotilla of warships to leave Stabiae? Was it on 26 August, the same day as the others found the Elder Pliny's body on the beach? Did they take the body straight to Misenum? Was the Elder Pliny the only casualty of those people who were cut off with him at Stabiae?

These are the questions which we cannot answer because they are not part of Pliny's story: but students may find it interesting to consider the possibilities.

It is worth pointing out that Pliny's account confirms archaeological evidence in one important respect. Herculaneum, the town nearest to the mountain, was engulfed in a flow of volcanic mud; but it was Pompeii, further south, which was covered with a fall of ash and stones. This suggests a strong north wind, which is precisely what Pliny narrates here. His uncle was able to sail eastwards with the wind on his left, then to turn southwards 'secundissimo [vento]' (line 24) towards Stabiae — but once having arrived there is unable to leave.

Finally there is the question of Pliny the Elder's behaviour. Was he being foolhardy, or courageous? Did he do all he could have done in the circumstances? It is possible that a full consideration of the circumstances may show him in a more favourable light than does a cursory reading of the letter.

Students could draw a map of the Bay of Naples as an aid to understanding Pliny's narrative.

Questions

1 Why was the Elder Pliny particularly interested in the phenomenon?

2 How did he come to have access to a fast boat?

3 What decision did the Younger Pliny make?

4 Why was Rectina in such danger?

5 In what words does Pliny express the change of purpose in his uncle's voyage?

6 What change did his uncle make in the kind of craft he ordered?

7 In what direction did he set sail?

8 What made the voyage more and more alarming?

9 Why were they prevented from landing at the intended place?

10 Where did the Elder Pliny decide to go instead? And for what purpose? What advice did he ignore? Who gave it?

11 What was preventing Tascius Pomponianus from escaping?

12 How did the Elder Pliny try to allay his fear?

13 How did he try to explain away the flames on the mountain to his friends?

14 How was it known that he had actually slept during the ensuing night?

15 What narrow escape did he have inside the villa?

16 What were the two dangers the company had to choose between? Which one did they choose? How did the Elder Pliny's attitude to the choice differ from that of his friends?

17 At what point did the Elder Pliny begin to feel ill? How did he show this?

18 Where was he when he died? Why did he die? What did his corpse look like when it was eventually found?

19 Do you think he was unlucky? In what way?

20 What impression do you get of the Elder Pliny's character?

21 What do you find the most vivid part of the account?

22 Do you find this an unhappy letter?

VI. 23

The letter

Pliny agrees to conduct a court case, on condition that the young
Cremutius Ruso acts as his junior.
For the letters in this Selection on patronage, see I. 19.
Nothing is known of Triarius.

Notes

2-3 **qui fieri potest . . . ?:** Pliny never charged fees for his services as a
barrister (V. 13*), following an old Republican tradition which had
lapsed with time. Claudius had legalised the taking of fees by fixing a
maximum of ten thousand sesterces (Tacitus, *Annals*, XI. 7).

4 **honestiorem:** 'which does me more credit . . .'

5 **Cremutius Ruso:** Pliny wrote IX. 19* to him, but he is otherwise
unknown.

6 **in:** 'in the case of . . . '

7 **adsignare:** an unusual word in this sense. In the gloss we have given
contextual meaning only.

8 **quod si cui:** 'but if I help anyone, I ought to help my friend Ruso . . . '

9 **mei:** objective Form D (genitive): 'his uncommon affection for me'.

11 **ante quam dicat:** Pliny asks Triarius to let Ruso take part in the case
without hearing him speak first: he must take Pliny's word that the
young man is good.

12-13 **spondeo . . . suffecturum:** a zeugma, because 'suffecturum' is used with
three slightly different meanings. Perhaps 'satisfy' is the nearest word
in English: 'I guarantee that he will satisfy your anxiety, my expectation and the importance of the case'.

15-16 **neque enim . . . emergere:** 'for no one has such outstanding ability
ready-made that he can become known . . . ' 'statim', which may be
translated 'ready-made', qualifies an 'est' which has been omitted.

16 **materia occasio:** asyndeton, frequent in Pliny.

Exploration

The final sentence puts this letter into its proper social perspective.
Patronage, as Pliny's Letters are constantly reminding us, was a central
feature of Roman life; and Pliny knew how much his own career, both
politically and in the courts, had been helped by patronage, particularly in its early stages. He had himself first appeared in the Centumviral Court at the age of eighteen (V. 8*) and had immediately done
well (I. 18*). But he tells us that in the old days not even the most
aristocratic young man could start his career in this court without being
introduced by a consular ('ne nobilissimis quidem adulescentibus locus
erat nisi aliquo consulari producente': II. 14*); and, though he is
referring to before his time, he may be taken to indicate that someone
very influential started him on his own legal career, especially as in this

same letter he deplores the young men now appearing in the Centum-viral Court.

In this present letter, Pliny is trying to get Cremutius Ruso off to a good start in his legal career, just as he himself had been helped. He says he has done this for others before (lines 5-6), no doubt having in mind Pedanius Fuscus and Ummidius Quadratus (VI. 11*), among others. Triarius would be unlikely to find Pliny's request in any way out of the ordinary. He would know that Pliny would not be foolish enough to want someone incompetent as his junior — after all, if Cremutius Ruso were to make a fool of himself, he would be making a fool of Pliny too.

Cremutius would be delighted to have this help from Pliny, who was perhaps by now the doyen of the Centumviral Court and may have drawn large crowds. And what better way for Cremutius to come to the notice of the public than by appearing with Pliny?

Questions

1 What has Triarius requested of Pliny?
2 Is anything revealed here about the fees which Pliny charged for pleading his cases?
3 What 'fee' does he say he is asking on this occasion?
4 Is it unusual for him to be helping a young man in this way?
5 Why does he consider this young man particularly suitable?
6 What long-term advantages does Pliny say will result from this patronage?
7 Does Pliny think patronage is essential for a young man start-ing off as a barrister?
8 What are the arguments for Triarius' consenting to this arrange-ment?
9 Are there many occupations which one can learn only by actually doing them?
10 Why do you think Pliny took on court cases at all?

VI. 25

The letter

Pliny has been asked for help in the search for a missing person; but, remembering a similar incident, he fears the search will be in vain.

Perhaps the recipient is Baebius Hispanus, to whom Pliny wrote I. 24*.

Notes

1 **splendidum:** almost a formulaic epithet for an *eques.*
2 **Ocriculum:** a town north-east of Rome on the Via Flaminia, about a day's journey from the capital.
4 **in aliqua inquisitionis vestigia:** 'on to some tracks in the search'.
7 **huic ego ordinem impetraveram:** men could be appointed centurion by the emperor after recommendation by an army commander or some other influential person. Metilius Crispus had clearly received his promotion through Pliny's exertions. The alternative route to becoming centurion was by promotion from lower ranks. See 'Exploration' for an *optio* who was in line for such promotion.
8 **proficiscenti:** i.e. to Rome, to receive his official appointment.
   quadraginta milia nummum: a generous donation, well in excess of what was required (S-W, 385); but Crispus was a fellow townsman of Comum. See I. 19 for Pliny's generosity to another such townsman, Romatius Firmus.
10 **a suis:** the meaning is 'a servis suis'.
13 **tuis:** *sc.* 'precibus'.
16 **cum quo fuisset:** i.e. Scaurus. 'May the gods show him their favour so he may find the man himself, just as he has already found the person he was with.'

Exploration

This is a fascinating letter — even in the highly developed and organised Roman empire people could just disappear, not only in remote parts but in Italy. Travel was simply more hazardous in the ancient world than it is today. There are numerous examples of this, and students could be invited to make a list. Such examples might include the parable of the good Samaritan; St Paul's journey by sea to Rome; the capture of Julius Caesar by pirates (piracy was a great hazard in the ancient Mediterranean); the death of Clodius as a result of armed violence in broad daylight on the Via Appia; the shipwrecked sailor of Horace, *Odes* I. 28; the death at sea of an anonymous *optio*, commemorated in the following acephalous inscription:

> . . . optionis ad spem ordinis centuria Lucili Ingenui qui naufragio perit   situs est (*RIB* 544)

. . . an optio in hope of promotion, of the century of Lucilius
Ingenuus, who died in a shipwreck. He lies buried.

(This inscription is in the Grosvenor Museum at
Chester — see *CLC* Unit III, slide no. 15)

Since this anonymous *optio* was presumably drowned, this is not a
tombstone inscription; instead of having at the end the initials H.S.E.
(hic situs est: he lies here) the 'hic' has been omitted and a space left
in case the body was found.

It is suggested that this inscription is read in connection with this
letter: it provides a suitable link between the general hazards of travel
in the ancient world and the particular disappearance here of the
centurion-elect Metilius Crispus.

Someone may well ask of Metilius Crispus: 'Perhaps he disappeared
because he chose to abscond with the 40,000 sesterces rather than
become a centurion?' This would be unlikely: a centurion was a respected man, received good pay and had prospects of promotion. Even in
the early principate, senior centurions (*primipili*) became prefects of
small provincial areas (e.g. *ILS* 1349 / EJ 243 for Baebius Atticus); they
were often patrons and benefactors of towns (for instance, a *primopilus*,
M. Helvius Rufus Civica, built a set of baths for the people of his town
and Sextus Pedius Lusianus Hirrutus an amphitheatre: *ILS* 2637 / EJ
248; *ILS* 2689 / EJ 244).

As with so many letters of Pliny, this present one throws incidental
light on patronage and Pliny's generosity.

Nothing else is known of the persons mentioned in this letter.

For travel in Roman times, see Paoli, chapter XX.

Questions

1 Where was Robustus last seen, as far as is known?
2 Does Pliny think he will turn up?
3 Why does Pliny tell the story of Metilius Crispus?
4 What are the two explanations he puts forward for such a disappearance? Is there a third possibility which he does not
mention?
5 There was no official help which a Roman could get to find
a missing person. What help can one get nowadays?
6 Would recommended centurions have made good soldiers?

VI. 34

The letter

Pliny praises Maximus for putting on a *munus* in honour of his dead
wife.

The various people called Maximus who occur in Pliny's letters are
rather a tangle. Perhaps this letter and VII. 26 were written to the same
person (Valerius?) Maximus.

Notes

1 **munus:** the proper word for a gladiatorial or wild beast show, because
such shows were a 'duty' or an 'offering' to the spirits of the dead, and
the blood (human or animal) which was shed was a propitiation to the
gods of the Underworld. (Hence Pliny's comments, lines 4-5, 'atque
hoc ... debebatur'.) It was difficult to believe, however, that the
crowds who attended such shows in imperial Rome paid any attention
to their religious significance: the show was an exciting entertainment
and a source of prestige to the person who arranged it. In Rome itself
this was nearly always the emperor.
Veronensibus nostris: Pliny had a strong sense of attachment to the
area where he was brought up, whether he was talking about his home-
town Comum and its surroundings (see II. 8, 'ad Larium nostrum') or
about a more extensive part of Transpadane Italy. Syme (*Tacitus*, I. 86)
defines this area, pointing out that it did not include all Transpadane
Italy, Verona was however included, so he associates himself with
Maximus who came from there.

4-5 **cuius memoriae ... :** 'to whose memory was owed either some public
building or a show — especially this kind of show, which is especially
appropriate for a funeral offering'. 'debebatur' is used to go with both
'memoriae' and 'funeri'.

7 **illud quoque egregie:** *sc.* 'fecisti'.

8 **per haec etiam:** 'even by things like this': Pliny seems to hint at a
personal lack of enthusiasm for such shows.

11-12 **tu tamen meruisti ... :** 'you nonetheless deserved that the credit should
go to you, because it was not your fault that you could not show
them'. The word 'acceptum' is a financial term, meaning 'entered in
the credit column'.

Exploration

It comes as a surprise to find Pliny praising Maximus for honouring his
dead wife with a gladiatorial show. Pliny might have been expected to
disapprove of gladiatorial shows on more than one count. To judge by
his dislike of the chariot-races in the Circus Maximus (IX. 6), he should
have disliked gladiatorial shows for much the same reasons. Then his
humanitas, demonstrated by his attitude to slaves (VIII. 16) and by his
tolerance towards condemned men turned public slaves (X. 31), should

have ranged him against the cruelty of a gladiatorial combat. But was
Pliny approving of the gladiatorial show? Surely he was approving of
Maximus discharging an obligation to the people of Verona in memory
of his wife. Perhaps he would have preferred him to discharge his obligation by means of an *opus* (line 4); but by the time Pliny wrote this
letter it was too late, the gladiatorial show had been staged.

Pliny seems to have been kindly disposed towards Maximus, perhaps
because they both came from the same part of Italy ('Veronensibus
nostris': line 1). Sherwin-White attributes meanness to Maximus, who
was apparently reluctant to put on a show because of the expense involved (S-W, 401). The people of Verona had perhaps been critical
because the panthers had failed to turn up. But Pliny rushes to
Maximus' defence and even calls him 'liberalis' (line 7). What would
he have said if Regulus (IV. 2) had put on a gladiatorial show in similar circumstances?

For dislike of gladiatorial shows, see Seneca, *Letters*, VII.

For modern accounts:

Michael Grant, *Gladiators* (Penguin, 1971).

J. P. V. D. Balsdon, *Life and leisure in ancient Rome*.

CLC Handbook to Unit I, pp. 88-9, cf. slides nos. 50-4.

Questions

1 What happened at a *munus*?
2 Why had Maximus put one on at Verona?
3 In whose honour was it?
4 Why was it a particularly suitable tribute?
5 Do you think Pliny really approved of a gladiatorial show like
   this? Do you find his apparent approval surprising? Was Pliny
   a religious man?
6 What word may suggest that Maximus himself disapproved of
   such shows?

VII.5

See commentary on VI. 7.

VII.16

The letter

Pliny has asked Calestrius Tiro to turn aside from his journey and perform a formal manumission ceremony for some of Fabatus' slaves.

For the five letters in this Selection concerned with Pliny's *humanitas*, see V. 19.

For L. Calpurnius Fabatus, see IV. 1.

Notes

1 **Calestrium Tironem:** this present letter tells us all we know about Tiro's political career. Pliny wrote four letters to him (I. 12*; VI. 1*; VI. 22*; IX. 5*).

2 **implicitum:** goes with 'Calestrium Tironem' (line 1); 'associated as he has been with me ...'
 simul militavimus: i.e. they were *tribuni militum* together. For a senator's career, see Introduction I: Background to Pliny's career, and II. 9.

3 **simul quaestores Caesaris fuimus:** in AD 90. Quaestors of the emperor were elected without contest and their duties included reading out the emperor's speeches and letters to the senate. For quaestors, see II. 9, note on lines 4-5.

3-4 **ille me in tribunatu ... praecessit:** Tiro was tribune in AD 91, Pliny in AD 92. Tiro was allowed to stand early under the terms of the *Lex Iulia de maritandis ordinibus* which allowed a remission of as many years between magistracies as a man had children. Perhaps remission could be granted only once for each child (S-W, 420). For the *ius trium liberorum* see X. 94.

4 **in praetura:** they were praetors together. Sherwin-White (pp. 763-71) and Syme (*Tacitus*, I. 76) agree on AD 93 as the year, but there has been much disagreement. After the praetorship Pliny's career was much swifter than Tiro's: Pliny was consul in AD 100 but Tiro had not been consul at the time of this letter, AD 107. See VI. 22* and IX. 5* for Pliny's advice and congratulation to Tiro as governor of Baetica.

5 **annum remisisset:** Pliny had already found enough favour to be a *quaestor Caesaris* and then Domitian allowed him to stand for the praetorship a year early. This is striking evidence for Domitian's backing of Pliny.

7 **provinciam Baeticam:** the province of (Hispania) Baetica comprised the southern part of the Spanish peninsula. It was a public province, i.e. one within the senate's jurisdiction. In 27 BC the Roman empire had been divided into *public* and *imperial* provinces. The governor of a public province was chosen *sorte* from a number of eligible candidates: he was styled *proconsul* (regardless of whether he was an ex-consul or

an ex-praetor) and he normally stayed only a year in his province. Imperial provinces were the responsibility of the emperor, who appointed the governors himself and kept them in their provinces for as long as he chose: he styled them *legati Augusti pro praetore.* The most important provinces, whether public or imperial, went to men who had actually held the consulship, the others to ex-praetors. (Egypt and certain provincial districts such as the Alpes Maritimae were given to *equites.*) A proconsul could perform non-contentious judicial acts when going out to, or returning from, his province (Ulpian, *Digest,* I. 16. 2), as this letter also shows. For a full discussion of Roman provincial government, see Fergus Millar, 'The Emperor, the senate and the provinces'.

9 **vindicta:** Form E (ablative): 'with a formal manumission ceremony'.
9-10 **vindicta liberare . . . manumisisti:** the various informal methods of manumission (see Paoli, 126), one of which Fabatus had already used, gave slaves the intermediate status of *Junian Latins.* To gain all the possible rights of freedmen, they needed to be manumitted formally *per vindictam* by a magistrate, and this is what Tiro was going to do. This would give them, among other things, the precious right of *conubium* so that their children would be free-born Roman citizens. See J. A. Crook, *Law and life of Rome,* 42ff.
10 **nihil est quod:** 'there is no reason why'
12 **quid velis:** an indirect question: 'ask yourself what your wishes are'.
13-14 **illi tam iucundum . . . :** an elliptical sentence that can be explained by adding 'est' after 'iucundum' and 'iubeo' after 'ego': 'My bidding is as agreeable to him as yours is to me.'

Exploration

This letter looks like a counterpart to IV. 1, in which Pliny explains to Fabatus that he and Calpurnia must turn aside from their journey to him to attend the dedication of a temple — public duty before private affection. Here now Tiro is to be asked to turn aside from his journey out to Baetica as its new governor to manumit some of Fabatus' slaves — a private matter before public duty. Note the language: 'deflectemus in Tuscos' (IV. 1); 'ex itinere deflectat' (VII. 16). Also, Fabatus is the recipient of both letters.

The letter seems to have been revised for publication. Though Pliny is an economical writer, and he and Fabatus are relatives, he feels it necessary to go into some detail about his and Tiro's parallel careers fifteen and more years previously. This must have been for the eyes of the world rather than of Fabatus. Why are such details given? The most likely explanation is that Pliny wanted by means of this letter to perpetuate the name and career of a friend. If so, he has succeeded, because all we know about Tiro comes from Pliny.

Questions

1 Why did Calestrius Tiro become tribune a year before Pliny did?
2 When did Pliny 'catch him up'? Why?

3 Where is Tiro now going? In what capacity? Is he choosing the
   most direct route?
4 What was involved in manumission?
5 What does Pliny say he will ask Tiro to do?
6 In what respect is Pliny having to reassure Fabatus?

VII. 21

The letter

Pliny, suffering from an eye complaint, assures Tertullus that he is looking after himself, and thanks him for his present of a chicken.

Cornutus Tertullus was Pliny's colleague both in the prefecture of
the Treasury of Saturn (AD 98-100) and as *consul suffectus* (AD 100).
For his career, see Appendix no. 2 and V. 14*. Tertullus came to the
consulship late (Syme, *Tacitus*, I. 82-3). Like Pliny, he was sent by
Trajan as a special envoy to Bithynia, probably being Pliny's immediate
successor there. It was a province with which he almost certainly had
connections. Pliny wrote one other letter to Tertullus, VII. 31*.

Notes

1 **collega carissime:** the informal tone suggests that Tertullus is not a
   'colleague' at the time of writing: he has been in the past (S-W, 428).
   infirmitati oculorum: eye diseases seem to have been extremely common in the times of the Roman empire, judging from the frequency
   with which they are mentioned. The general ignorance of hygiene
   encouraged them.
4 **difficulter sed abstineo:** an idiomatic ellipse, equivalent to 'difficulter
   abstineo, sed abstineo'.
6 **cryptoporticus:** a hybrid word, possibly invented by Pliny, to describe
   a *porticus* or colonnade whose arches had walls and windows between
   them. See II. 17*.
8-9 **balineum adsumo quia prodest, vinum quia non nocet, parcissime
   tamen:** compare II. 8, lines 6-7: 'qui sic concupisco ut aegri vinum
   balinea fontes'.
9-10 **et nunc custos adest:** Sherwin-White thinks the 'custos' is one of Pliny's
   private doctors (S-W, 429). But might it not be Pliny's wife?
11-12 **quam satis . . . vidi:** rearrangement is necessary to translate this sentence. 'Though I still have inflamed eyes, my sight was keen enough
   to see that it is a very plump one.' Students may find it easier to reach
   the meaning if the core of the sentence, 'quam pinguissimam vidi', is
   taken first.

Exploration

This is one of the letters which are 'close to the realities of correspondence' (S-W, 12). It comes across as informal and unaffected. It is partly to thank Tertullus for his present of a chicken and for the concern he is showing for Pliny's eye complaint, but it also shows, in the words 'hic non stilo modo verum etiam lectionibus difficulter sed abstineo' (lines 3-4), the importance Pliny attached to his literary pursuits. For this, see also I. 6, etc.

Questions

1 What is Pliny suffering from?
2 What restriction is he having to observe as a result?
3 Explain what he means by 'solis auribus studeo' (line 4).
4 Who do you think was the 'custos' (line 9)?
5 Why do you think this particular complaint was so common in ancient Rome?

VII. 24

The letter

Pliny writes anecdotally about Ummidia Quadratilla, who has just died at an advanced age, and in laudatory terms about her grandson, a promising lawyer.

Rosianus Geminus, to whom four other letters were written, was a protégé of Pliny. He was Pliny's consular quaestor in AD 100, and in X. 26* Pliny recommended him to the Emperor Trajan. For his career, see S-W, 402. For consular quaestors, see II. 9, note on lines 4-5.

For the 'pen-portrait' letters in this Selection, see V. 5.

Notes

1 **Ummidia Quadratilla:** this lady came of a well-to-do, consular family. Her father, C. Ummidius Durmius Quadratus, became a senator in AD 14 and served all five of the first Roman emperors (*ILS* 972 / Smallwood: (1) 229). He died in AD 60 as governor of Syria, being unfavourably contrasted by Tacitus with one of his heroes, Corbulo (*Annals,* XIII. 8, for instance). Ummidia herself was born about AD 26, sharing her father's longevity. The family came from Casinum, where Ummidia built at her own expense an amphitheatre and temple for the people: 'Ummidia C.f. Quadratilla amphitheatrum et templum Casinatibus sua pecunia fecit' (*ILS* 5628); cf. *CLC* Unit III, Stage 28 and *Handbook* pp. 113f.

4 **honestissimo testamento:** a Form E (ablative) absolute, if it is possible
   to classify a phrase such as this. Compare 'decessit veteri testamento'
   (V. 5, line 5). The will is 'honestissimum', because Quadratilla left
   her estate to her family (lines 4-5). Pliny returns to the point in line 33,
   when he refers to Quadratilla's 'pietate'.

5 **nepotem:** C. Ummidius Quadratus, a friend of Pliny and the recipient
   of two letters from him (VI. 29*; IX. 13*). In VI. 11* Pliny praises his
   promise in the courts (see 'Exploration'). Surprisingly little is known
   about him. But he was *consul suffectus* in AD 118, falling into dis-
   favour late in Hadrian's reign (S-W, 431). He was not in the male line
   and must have taken the family names by adoption — perhaps to
   honour his grandmother, while she was alive, or on inheriting at her
   death?

5-8 **neptem parum ... amandum:** 'I know the granddaughter all too little;
   the grandson is a very dear friend of mine — a remarkable young man
   and one who inspires in more people than his actual kin such love as
   puts him among one's relatives.' The chief difficulty is caused by 'inter
   propinquos amandum', which expresses the idea 'to be loved along with
   one's relatives'.

8 **ac primum conspicuus forma:** 'in the first place, though noticeable for
   his good looks'.

9-10 **intra quartum et vicensimum annum maritus:** 'getting married before
   the age of twenty-four' — i.e. he escaped the disadvantages which the
   Lex Iulia imposed on people who were still unmarried after the age of
   twenty-five, the most punishing of these being that a bachelor over that
   age was not allowed to receive any inheritance or legacy.

11 **delicatae severissime:** there is a nice contrast here between Ummidia
   Quadratilla and the way of life of her grandson.

12 **pantomimos:** for *pantomimi*, see Carcopino, 243ff; *CLC Handbook*
   for Unit III, pp. 114-16. The *pantomimus* C. Ummidius Actius recorded
   on an inscription at Puteoli (*ILS* 5183) may perhaps be one of
   Ummidia Quadratilla's company.

15 **audivi ipsam:** *sc.* 'dicentem'.

16 **ut feminam in illo otio sexus:** 'being a woman, with the usual leisure
   of her sex'. Pliny's generalisation must of course have applied only to
   wealthy women, who were relieved of all domestic chores by their
   slaves and could not pursue public careers.

19-20 **quod mihi ... videbatur:** 'non ... magis ... quam' is difficult, as perhaps is the idea of the elderly Quadratilla showing respect for her grandson. 'amore' and 'reverentia' are Form Es (ablatives) of cause: 'out of
   love ... out of respect'. In translation 'reverentia' may best be placed
   before 'amore': 'and she seemed to me to be doing this no less out of
   respect than out of love for him'.

21-2 **proximis sacerdotalibus ludis:** dramatic shows are clearly meant here;
   but 'the phrase is too loose for exact dating' (S-W, 432). There were
   some 150 days of *ludi* from April till November.

22 **productis in commissione pantomimis:** Quadratilla's freedman had been
   entered for an exhibition performance — hence the *clientes* acting as
   cheer-leaders (see below, lines 25ff).

24-5 **hoc nepos:** *sc.* 'dixit'.

25 **alienissimi homines:** by what follows Pliny makes it clear that these are
   clientes of Ummidia Quadratilla. The juxtaposition of 'nepos' (the rela-
   tion who rightly received an inheritance) and 'alienissimi' (who had no
   such right and remained outsiders as a result of Quadratilla's will) is
   worth noting.

28 **singulos gestus dominae cum canticis reddebant:** 'gestus' can be used of
   a gesture or mannerism of an actor: so Cicero speaks to Atticus 'de gestu
   histrionis' (*Ad Atticum* VI. 1. 8). 'canticum' was a song accompanied
   by music and dancing: 'ac desaltato cantico abiit' (Suetonius, *Gaius
   Caligula*, 54). Perhaps Quadratilla was imitating the gestures of her
   pantomimus, as well as singing to his dancing, and the *alienissimi*
   joined in; or perhaps they were just aping her gestures of approval or
   disapproval.

29 **exiguissima legata ... corollarium:** these people had hoped to receive
   legacies in Quadratilla's will, but all they got was a gratuity ('corollarium') for applauding in the theatre. The words 'theatralis operae'
   suggest that they are a claque of applauders. So Percennius, who played
   some part in the army mutinies of AD 14, was described by Tacitus as
   'dux olim theatralium operarum' (*Annals*, I. 16).
   For legacy-hunting, see *Exploration* below. Pliny evidently held it in
   contempt: see, for instance, II. 20* sections 1-8 in *CLC* Unit IV 'domi',
   which concerns the legacy-hunting activities of Pliny's bête-noire
   Regulus.

30 **herede:** there is a pointed contrast between this word and 'exiguissima
   legata' (line 29).
   haec: *sc.* 'scripsi'.

32 **quod ... retractare:** 'scribendo' goes with 'retractare', whose object is
   'quod ceperam gaudium'.

34 **C. Cassi:** in the early imperial period there were two 'schools' of law,
   the Cassian and the Proculian. C. Cassius Longinus was *consul suffectus*
   in AD 30 and subsequently governor of Asia and Syria. Nero exiled
   him in AD 65 and he was recalled by Vespasian. In the debate on
   whether or not to execute the whole household of slaves of the murdered Prefect of the City, Pedanius Secundus, Cassius took a tough
   line (Tacitus, *Annals*, XIV. 43-4). His severity is mentioned earlier with
   regard to disturbances at Puteoli (*Annals*, XIII. 48), the two references
   perhaps supporting the view that the Cassian 'school' of law was 'con-
   servative, rational, republican and attached to principles' (S-W, 433-4;
   Syme, *Tacitus*, I. 447-8).

37-8 **cum tantus orator ... :** 'when he goes forth from it as great an orator
   as Cassius was a jurist'.

Exploration

Ummidia Quadratilla is one of the most colourful people to appear in
Pliny's *Letters*, perhaps being rivalled only by Regulus (for whom, see
IV. 2, for instance). The comparison with Regulus is informative.
Pliny's aversion to Regulus is clear, and he is (as this present letter
shows) not entirely approving in his attitude to Ummidia Quadratilla.
He approves of her for leaving an honourable will (line 4), for the way
she sent her grandson off to his work, while she amused herself playing
draughts or watching her *pantomimi* (lines 15-20) and for her family
affection (line 33). But he considered it unbefitting a great lady to have
a company of *pantomimi* and pander to them (lines 12-13). One is
tempted to feel that Pliny, with his crowded public career and consuming intellectual interests, could not enter into the predicament of
this wealthy and bored old lady.

No one can accuse Pliny of being an artless writer, and it is easy to guess that his equivocal portrayal of Ummidia Quadratilla is meant as a foil to his pen-portrait of the grandson. Even if he wanted to, Pliny could not be wholly critical of the grandmother — after all, Ummidius Quadratus was her grandson, had lived in her house and no doubt retained great affection for her. (Would anyone as exemplary as Ummidius Quadratus do otherwise than feel warmly for his grandmother?)

Why did Pliny think so well of the grandson? This letter shows that he had had a hand in his education (line 15). Another, VI. 11*, speaks glowingly of the promise which Ummidius Quadratus and another young man, Fuscus Salinator, were showing in the courts, Pliny adding of the two men: 'me aemulari, meis instare vestigiis videbantur'. Pliny could hardly castigate such men. Instead he ends this present letter with a gracious compliment: Quadratus is to live in the house once owned by Cassius, and Pliny believes that he will become as great an orator as Cassius was a jurist.

Students will have a chance, through reading this letter, to discuss the two people depicted here, Ummidia Quadratilla and her grandson. The grandson is, as Pliny writes him up, a paragon of the virtues and perhaps for this reason less attractive and less credible than his grandmother, a lady who has outlived most of her contemporaries and who whiles away her hours with pastimes which the serious-minded Pliny finds indecorous and frivolous. Use could be made of the inscription recording Quadratilla's benefactions to Casinum (*ILS* 5628, quoted above in the first note).

There is also a chance to discuss legacy-hunting, a popular Roman pastime among the rich and, as this letter suggests (lines 25-30), among the not-so-rich. (Rich people would have acted with more outward decorum than to behave so outrageously in a public place.) These particular legacy-hunters received only a small reward for their pains. Students should read VIII. 18* for an account of the will of Domitius Tullus, who gave appropriate bequests to his family but left the legacy-hunters disappointed — the city was in an uproar of conflicting opinions following his action. Not surprisingly, legacy-hunting is a frequent topic in the *Epigrams* of Pliny's contemporary, Martial. Here is one such epigram, stridently blunt:

> scis te captari, scis hunc qui captat, avarum,
>   et scis qui captat quid, Mariane, velit.
> tu tamen hunc tabulis heredem, stulte, supremis
>   scribis et esse tuo vis, furiose, loco.
> 'munera magna tamen misit.' sed misit in hamo;
>   et piscatorem piscis amare potest?
> hicine deflebit vero tua fata dolore?
>   si cupis, ut ploret, des, Mariane, nihil.

(VI. 63)

Horace devotes *Satires*, II. 5 to the topic of legacy-hunters; cf. *CLC* Unit
IV 'domi' and *Handbook* pp. 129f.

Questions

1 How old was Ummidia Quadratilla when she died?
2 What kind of bodily constitution did Ummidia Quadratilla
   have?
3 Can you suggest why an older married woman might *not* have
   had 'corpus compactum et robustum', especially in ancient
   Rome?
4 What kind of will would Pliny *not* have described as 'honestissimo'?
5 How does Pliny try to convey the young man's lovable qualities?
6 What kind of scandal might have attached to a very handsome
   young man?
7 Why was it a particularly good thing for him to be married by
   the time he was twenty-four? How would it have affected his
   grandmother's will if he had not done so?
8 What attitude did Quadratilla take towards her grandson's
   pursuits?
9 Does Pliny approve of their relationship?
10 Why did some women have so much *otium*?
11 What surprising remark had Quadratus made to Pliny?
12 What are Pliny's feelings towards the *clientes* of Quadratilla
   in the theatre? What individual words make these feelings
   clear?
13 How are the *clientes* now being rewarded? What is the irony in
   this?
14 Why is Pliny pleased about the whole thing?
15 What was special about Quadratus' house? What is going to
   happen to it now?
16 Does Pliny like the young Quadratus because he is like himself?
17 Can you think of any times when Pliny himself preferred to
   study although there was something exciting happening
   (see VI. 16; IX. 6)?
18 Who is this letter really about, Ummidia Quadratilla or
   Quadratus?

VII. 26

The letter

Pliny reflects on the changed feelings of people who are ill.
For the problem of who Maximus was, see VI. 34.

Notes

3-4   **non adpetit honores:** among the senatorial class, and among the *equites* too, there was a constant striving after promotion. Note Pliny's attempt to advance the career of Sextus Erucius (II. 9).

5   **satis habet:** 'regards as enough'.

   tunc deos . . . meminit: in a simpler form this sentence might read, 'tunc deos esse meminit, tunc se esse hominem meminit'.

9-10   **mollemque . . . pinguem . . . innoxiam beatamque:** all these adjectives go with 'vitam'.

10-14   **possum ergo . . . infirmi:** this sentence may become clearer if the relative clause 'quod . . . conantur' is translated after 'praecipere'. The final 'ut tales . . . ' can become a new sentence. An alternative method, which better retains the emphasis of the Latin, is simply to postpone 'possum' till after 'ipse'.

Exploration

This letter is not so much about illness as about the profounder reflections which it provokes. For Pliny, people are at their best when they are ill: they are prevented from pursuing a great many normal activities which he sees as unpleasant; they come face to face with their mortality and they make resolutions to become better people when they recover. Among other things, a sick man takes no interest in malicious gossip (lines 6-7); and there is an obvious parallel here between this letter and I. 9, where Pliny, in the tranquillity of his villa at Laurentum, notes that there he is away from malicious talk.

These two letters have more general parallels: at his Laurentine villa Pliny has temporarily escaped the demands of public life, while a sick man has also (involuntarily) escaped them. Here is something which disturbs Pliny profoundly: there is in him the constant struggle between the attractions of a quiet life and the supreme reward of an ardous public career, which is the chance to gain immortality (IX. 3*). Pliny never resolves this struggle to his complete satisfaction, and in the present letter he clearly approves of the good resolutions of a sick man, regarding the 'mollis et pinguis vita' as a morally superior life, since it is 'innoxia et beata' (lines 9-10).

It is interesting that Pliny sees sick people as 'optimos' because they are not doing certain things such as pursuing honours or wealth (lines 3-5). For a Roman, goodness so often meant merely the absence of

wrongdoing, rather than anything positive, and herein lay the attraction
of a passive philosophy such as Stoicism.

Questions

1 What does Pliny say is the invalid's frame of mind?
2 Do you agree with his description?
3 What is the important question which Pliny says he can
   answer?
4 Think of a title for this letter.
5 Does the letter suggest that Pliny disliked certain aspects of
   himself?

VII. 29

The letter

Pliny is outraged by the monument of the long-dead freedman, Pallas,
who became Claudius' financial secretary.

It is not possible to identify Montanus, who also received a second,
much longer letter on the subject of this monument to Pallas (VIII. 6*).

Notes

1 **legeris:** future perfect. The object of this verb is an omitted antecedent
   to 'quod': ' . . . if you read something which you cannot believe unless
   you have read it'.
2 **via Tiburtina:** Form E (ablative) of place: the road to Tibur (modern
   Tivoli), eighteen miles east-north-east of Rome. People were buried by
   the sides of roads leading out of Rome. Burial within the walls of the
   city was prohibited, except for the imperial family.
3 **Pallantis:** Tacitus related the events surrounding the bestowal by the
   senate of this honour on Pallas. Pallas was the freedman of Claudius'
   mother, Antonia. He became Claudius' *a rationibus* (financial secretary). He was one of those responsible for Agrippina the Younger's
   marriage to Claudius and he also helped to secure Claudius' adoption
   of her son Nero. But Nero as Emperor forced Pallas to retire (AD 55)
   and then had him put to death in AD 62 in view of his great wealth
   (Tacitus, *Annals,* XIII. 14; XIV. 65). For the power and influence of
   Claudius' freedmen, see Fergus Millar, 'Emperors at work'.
4-5 **patronos:** i.e. the imperial family, whose freedman he was.
5 **ornamenta praetoria:** the senate granted him the rank of a praetor,
   although he was a freedman.
6 **cuius honore contentus fuit:** i.e. Pallas accepted the honour but refused
   the money. 'cuius' refers back collectively to the whole of the senate's
   decree: 'but of all this he accepted the decoration only'.
7 **quae . . . :** 'things which . . . ' Pliny is referring to titular honours rather
   than the offices of the *cursus honorum.*

a iudicio: this phrase can be taken (as Pliny probably intends) in two
different ways: 'from human decision' or 'from good judgement'.

9-10 caenum ... sordes ... furcifer: the most violent words that Pliny uses
of anybody.

11-12 ut moderationis exemplum posteris prodere: i.e. by having the stone
inscribed. This gibe would have a point only if Pliny knew that Pallas
himself had ordered the inscription; but he does say that he has only
just noticed the monument, and it is only in the later letter (VIII.
6*) that he says he has found out the full story of the senate's decree. Who
did order the inscription we do not know, nor did Pliny. This is the
only place where he descends to this kind of uninformed abuse.

12-14 ridere satius ... ut rideantur: 'It is better to laugh, lest they think they
have achieved something great, these people who by good luck reach
such a position only to become a laughing-stock.'

13 huc: 'to such a position'.

Exploration

Towards the end of the letter Pliny puts a brave face on it and comments: 'But why am I getting angry? Better laugh instead.' But he pro-
tests too much that laughter, rather than anger, is the appropriate
emotion. He is angry — otherwise why did he write not just this short
letter but another much longer one (VIII. 6*) to Montanus on this
same subject?

Why is he angry about a man who received these honours and died
half a century or so earlier? For Pliny, Pallas' monument, set up in-
decently near the city walls ('intra primum lapidem'; lines 2-3), was an
uncomfortable reminder that under the principate the senators, the old
Republican rulers of Rome, had a diminished status little better than
that of administrators, and that upstarts (and foreign upstarts at that,
Pallas being a Greek) could reach praetorian rank by backstairs influence
in the palace. It took Pliny years of toil up the *cursus honorum* to
reach his position; and though he was now a *consularis*, while Pallas
only gained praetorian status, Pliny knew who was the more influential
in his own day. Was it all worth it to leave the deep peace of Laurentum
or the lakeside of Comum and the precious company of his books, if a
freedman could win himself such influence and be honoured by a servile
senate?

It is interesting to compare Pliny's two letters about Pallas, shrill and
almost hysterical, with Tacitus' succinct condemnation of this same
senatorial decree, by which 'libertinus sestertii ter miliens possessor
antiquae parsimoniae laudibus cumulabatur' (*Annals*, XII. 53).

Questions

1 What feelings does Pliny say Montanus will have on hearing
what follows?

2 What has Pliny seen?

3 Had Pallas been dead a long time?

4 What did the monument say had been bestowed on him?

5 What had he refused?

6 Explain the words 'quae saepius a fortuna quam a iudicio proficiscerentur' (lines 7-8).

7 What does Pliny think of Pallas?

8 Is there anything here you would regard as unfair comment? Is there any criticism which may be completely without foundation?

9 Is Pliny annoyed with himself?

10 Explain the reasons for Pliny's attitude towards Pallas.

11 What does Pliny say is the proper reaction to what he has seen? Is he right? How successfully does he follow his own advice?

VIII.16

The letter

Pliny, grieving at illness and deaths among his slaves, is consoled by the thought that he has treated them well.

For Pliny's humane attitude towards slaves and freedmen, see list in notes on V. 19. See also Seneca, *Letters* XLVII.

C. Plinius Paternus is probably the Paternus who received IX. 27. If so, Pliny wrote four letters to this Paternus, who may well have come from Comum (S-W, 135).

Notes

2 **solacia duo:** *sc.* 'sunt'.

3 **facilitas manumittendi:** 'my readiness to give them freedom'. For the methods of manumission, see notes on VII. 16. See also Paoli 126. The prospect of dying a free man would be a real consolation to a slave who was sick.

5-6 **quasi testamenta facere:** this was apparently beyond contemporary custom (S-W, 467) and was not legally valid: indeed slaves could not even own property in the eyes of the law.

7 **iussus:** Form A (nominative) of perfect participle passive. Pliny acts as if he were a proper legal executor.

7-8 **dumtaxat intra domum:** bequests to slaves in other households would be difficult if the other master did not have Pliny's liberal habit of allowing slaves to have property.

12 **alios:** 'other people' — the subject of 'vocare', whereas 'casus' is the object.

13 **eoque:** 'and for that reason . . . '

15 **hominis est:** this is the *characteristic* Form D (genitive).

17 **plura:** *sc.* 'dixi'.

19 **apud quem:** this can almost be translated as 'from whom'.

Exploration

Any discussion of this letter must centre around the treatment of slaves, and reference to other letters and even other writers will be useful. For the rights of slaves, and for further references, see Paoli, chapter X. Students should note that in the care of his slaves Pliny went far beyond what was legally required of him. He is aware of this, and in the present letter is both congratulating himself on his own *humanitas* and criticising other, harsher masters. Sherwin-White even thinks that this letter is a mild rebuke to Paternus, who had once advised Pliny on the purchase of slaves (I. 21*); this rests on the assumption that Paternus is the wealthy but mean man who won himself a Martial epigram, XII. 53 (S-W, 135).

A more general discussion may arise about the *prima facie* incongruity of Pliny's recognising the human dignity of his slaves but still assenting to the possession of them. Is he merely behaving according to his social conditioning, or does he accept that the system cannot be changed and resolves to act within it as humanely as possible?

Questions

1 What has distressed Pliny?
2 Explain what he calls his 'two consolations'.
3 Could a slave legally make a will? Did a slave legally have possessions?
4 What does Pliny say is the reason for his distress? Is he ashamed of it?
5 How did some other people react to such happenings? How does Pliny describe those people?
6 How would you translate 'homines' (line 14)?
7 What is he expecting the feelings of Paternus to be?
8 What do you understand by the word 'manumittere'?
9 Does this letter strike you as sincere?

VIII.17

The letter

Pliny describes the severe flooding of the Tiber and the Anio, and asks Macrinus to let him know whether he is safe.

The recipient is probably Caecilius Macrinus, to whom Pliny wrote four other letters (II. 7*; VII. 6*; VII. 10*; IX. 4*).

Notes

1 **istic . . . hic:** we do not know where Macrinus was but Pliny was in or near Rome.
3 **fossa:** for the Emperor Trajan's *fossa*, see S-W, 468. We do not know where it was, but Pliny's silence about Rome itself suggests that it prevented flooding within the city.
4 **quam:** this relative pronoun refers to 'fossa'.
6 **quae solet flumina accipere:** 'the rivers which it normally receives': 'flumina' is the object of 'cogit' (line 7).
8 **Anio:** a seventy-five mile long tributary of the Tiber, which it joined just north of Rome.
13 **viderunt quos . . . :** 'the people who have been marooned on higher ground by this flood have seen . . . '
18 **illa:** *sc.* 'loca'.
24 **ne quid simile istic:** *sc.* 'acciderit'.
28-9 **doleas . . . timeas . . . :** these are potential subjunctives 'for you can suffer only what you know has happened, but you can fear everything that can possibly happen'.

Exploration

Pliny had a particular interest in rivers and their problems. For three years, AD 104-6, he was 'curator alvei Tiberis et riparum et cloacarum urbis' (Appendix no. 1; see also Introduction, II: Pliny's career, p. 8). His interest in the problems of water engineering stayed with him. For instance, he wrote from Bithynia to Trajan about an aqueduct for Nicomedia (X. 37), about a canal from the Lacus Sunonensis (X. 41; 61) and about an open sewer at Amastris (X. 98).

One of Pliny's strong points is his ability at simple but vivid description. This letter may be compared with the two accounts of the eruption of Vesuvius (VI. 16; VI. 20*), the dolphin letter (IX. 33) and the description of his villa by the sea at Laurentum (II. 17*).

Flooding, a common occurrence in Italy, is frequently described by ancient writers. Here from several possible examples is a description by Ovid:

> exspatiata ruunt per apertos flumina campos
> cumque satis arbusta simul pecudesque virosque
> tectaque cumque suis rapiunt penetralia sacris
> si qua domus mansit potuitque resistere tanto
> indeiecta malo, culmen tamen altior huius
> unda tegit.

(Metamorphoses, I. 285-90)

Students may find the sheer weight of vocabulary in this letter discouraging. Even the more able may find it advisable to go through the vocabulary before preparing the letter; they may find that some words could have been guessed (e.g. vallēs: vallis, f., valley), and certainly some will have been met before (e.g. rīpa).

Questions

1 Where is Pliny writing from?
2 Did he have any specialist knowledge of flooding problems?
3 What precaution does he say has been taken against this kind of thing? Why does it seem to have been partly unsuccessful?
4 What do you understand by 'adiacentibus villis velut invitatus retentusque' (line 9)? Is this a vivid picture? To what does it serve as a contrast?
5 What has happened in the parts the flood has not reached?
6 What is the literal meaning of 'deiecti' (line 19)?
7 Comment on the lines 19-23. Do you find that this clipped style, with the verbs omitted, is effective? Vivid?
8 What does Pliny ask Macrinus to do?
9 Put the maxim of the last sentence in your own words. Do you agree with it?
10 Do you find the description in general to be a vivid one? Which part do you find most evocative?
11 It was common for the emperor to give a relief grant for people who suffered this kind of disaster. What is usually done nowadays?
12 Does modern insurance cover disasters like this?

IX. 6

The letter

Pliny despises the Ludi Circenses and spends his time writing during the days they are held.

For Pliny's literary activities, see the references in notes on I. 6.

Calvisius Rufus was a town-councillor of Comum, Pliny's *patria* or home-town. He was a friend and business adviser of Pliny, who wrote half a dozen letters to him.

Notes

1-2 **iucundissima quiete**: Pliny expects that Calvisius will be surprised that he can find 'quies' in Rome. But then the city is quiet, because everyone is at the Circus Maximus.

3 **Circenses**: Chariot races were between jockeys (*aurigae, agitatores*) who belonged to four stables or riding-schools (*factiones*). (The number of stables had originally been two, had risen to four and had been increased to six by Domitian, but had settled down to four again.) Each stable had a colour: red (*russata*), white (*albata*), green (*prasina*) and blue (*veneta*). The jockeys wore shirts of the colour of their particular stable. One of the most famous jockeys was a Lusitanian Spaniard called Gaius Appuleius Diocles, who was born in AD 104 and died at the age of forty-two in Antoninus Pius' reign. His lengthy epitaph (*CIL* XIV. 2884, printed with a translation in Donald Dudley, *Urbs Roma,* 214-15) records *inter alia* that he won 1,462 victories and 35,863,120 sesterces in prize money. See also 'Exploration' for a charioteer's epitaph.

As this letter suggests, the popularity of the chariot races as a spectator sport seems to have been immense. Juvenal remarks:

> totam hodie Romam circus capit, et fragor aurem
> percutit, eventum viridis quo colligo panni.
> (*Satires* XI. 197-8)

Each race was of seven laps (about 5 miles) and lasted about a quarter of an hour. Racing went on for several days at a stretch.

The Emperor Caligula supported the green *factio* (Suetonius, *Gaius Caligula,* 55), as did Nero and Domitian. Vitellius supported the blue *factio.* Support for the *factiones* could be fanatical. Gibbon, writing of a later age, describes the way the blue *factio,* apparently backed by the Emperor Justinian, hounded supporters of the green *factio* out of Constantinople and how it was not safe to be in the streets of the city at night (*The decline and fall of the Roman Empire,* chapter XL).

For the Circus Maximus, which was situated between the Palatine and the Aventine hills, see (for instance) Donald Dudley, pp. 211-16. For discussions of the chariot-races, see Carcopino, pp. 234-43, Paoli, p. 251, Balsdon, *Life and leisure in ancient Rome,* pp. 314ff. and for illustrations see *CLC* Unit IV slides nos. 32-4.

3-4 **quo genere . . . teneor**: this may be translated as a coordinate clause: 'and I am not gripped by this type of spectacle even to the slightest degree'.

4 **nihil novum:** *sc.* 'est'.
  non: qualifies 'sufficiat', not 'semel'.
9 **favent panno pannum amant:** a chiasmus in word-order only, not in
  thought, as in lines 10 and 16-17.
17-19 **ac per hos dies . . . :** the difficulty in translating this is to express not
  only the play on 'otium' and 'otiosissimis' but also the oxymoron
  'otiosissimis occupationibus'.

Exploration

Ironically, a student's interest in this letter is likely to centre on the
chariot-racing. Some will have seen *Ben Hur* and its chariot-race, while
other film-goers may remember the remark in *Junior Bonner,* 'When
you have seen one rodeo, you have seen them all'. Discussion may most
easily start with chariot-racing. The Diocles inscription is too long to be
given here, and too technical to translate easily. Instead students could
be invited to translate and discuss *ILS* 5282:

> dis manibus Epaphroditus, agitator factionis russatae, vicit
> CLXXVIII, et at (=ad) purpuream libertus vicit VIII. Beia
> Felicula fecit coniugi suo bene merenti.

(*Purpurea* and *aurata* were the two extra *factiones* added by Domitian.)
  The structure of the letter is interesting. There is the obvious contrast between the crowds and the activity at the races and the atmosphere of quiet which surrounds Pliny among his writing-tablets. There
is the contrast too between Pliny's disparagement of the races and his
devotion to literary pursuits. It would be wrong to see the structure of
this letter as presenting Pliny away from things on the edge, while the
great mass of the people is in the middle at the Circus Maximus. Rather,
the return at the end from the chariot-races to Pliny and his writings
gives the letter unity and the reader an added appreciation of Pliny's
theme.

Questions

1 How has Pliny been spending his time?
2 Why has there been nothing else to do?
3 Did any public business go on at times like these?
4 What word does he use to describe the mentality of people
  who watched the games?
5 How does he illustrate that the people's favour towards the
  charioteers meant nothing at all?
6 Does any phrase seem quite unashamedly snobbish?
7 What dismays Pliny most of all?
8 Explain the word-plays in the last sentence (see note above).

9 Is Pliny missing something? What do you think was the attraction of these events? Does his comment about the colours miss
  the point?

10 Why does he not mention the gambling, when this was a major
   part of the proceedings for many people?

11 Do you admire Pliny for having the courage of his convictions,
   when many people whose esteem he desired — including the
   emperor — did watch the races?

12 What does this letter tell us about Pliny's attitude to life? Does
   it agree with impressions gained from other letters?

IX . 12

The letter

Pliny uses a personal anecdote to advise Junior not to be too strict with
his son.

The recipient is probably C. Terentius Junior, an equestrian and a
landowner at Perugia. He decided not to accept Domitian's offer of
senatorial rank and lived quietly in the country.

Notes

2 **equos et canes**: note IV. 2, where Regulus' son is said to have had many
  Gallic ponies and large and small dogs (as well as nightingales, parrots
  and blackbirds!).
  huic ego: *sc.* 'dixi'.
  heus tu: a colloquial phrase, addressed only to an intimate or a person
  of lower social standing.

4-5 **si repente pater ille**: *sc.* 'fieret': 'ille' is the subject, 'pater' the complement.

7 **haec**: object of 'scripsi' in the next line.

10 **atque ita hoc . . .** : 'and use this fact of being a father in such a way
   that . . . '

Exploration

Pliny seems to use the anecdote of the over-strict father as a gentle
introduction to the real purpose of the letter, his advice to Junior to be
more tolerant of his son.

Some may find Pliny interfering and sanctimonious here, but it must
be remembered that he had no children of his own and perhaps reacted
strongly when he thought a father was taking his son for granted. A
cynic might retort that it is easier to be humane in theory than in practice, but Pliny does not always adopt a permissive attitude, saying this

of the young of his day: 'statim sapiunt, statim sciunt omnia, neminem verentur, imitantur neminem atque ipsi sibi exempla sunt' (VII. 23*).

Individual young people he is ready to help or to praise: Cremutius Ruso (whom he wants to act with him in court — VI. 23), Pedanius Fuscus and Ummidius Quadratus, though it must be admitted that these last two 'me aemulari, meis instare vestigiis videbantur' (VI. 11*).

Questions

1 Why was the father reproving his son?
2 How do you know that Pliny is on intimate terms with the father?
3 What arguments does Pliny use in reproving him?
4 Did Pliny think the son's behaviour was perfectly all right?
5 How would you translate the word 'homo' which occurs twice in the last line?
6 Is Pliny being foolishly permissive?

IX. 15

The letter

Pliny finds that all is not pleasant in the country, and asks for news of Rome.

Three other letters in the Selection concern Pliny in the country: I. 6; I. 9; II. 8.

The recipient, Q. Roscius Coelius Murena Pompeius Falco, received three other letters from Pliny (I. 23*; IV. 27*; VII. 22*). He had an impressive career. He was *consul suffectus* in AD 108, having *inter alia* been governor of Lower Moesia, Britain and Asia (for his career, see *ILS* 1035-7).

Notes

1 **in Tuscos:** these are the estates at Tifernum-on-Tiber, referred to in IV. 1. See V. 6* and IX. 36* for descriptions by Pliny of his house on his Tuscan estates and his praise of life there.
2 **at hoc ne in Tuscis quidem:** *sc.* 'facio'.
4 **invitus:** *sc.* 'lego'.
   retracto: i.e. for publication.
8-9 **hactenus ... quod ... percurro:** 'to the extent of riding around some part of the estate — for the exercise'.
9-10 **nobisque sic rusticis ...:** 'and while I am in this rustic state, write to me in full about your urbane pursuits'. It is not really possible to translate the two double meanings.

Exploration

This letter comes as something of a surprise after I. 9 (where Pliny talks
of the emptiness of city life and the joy of escaping to his villa at
Laurentum) and II. 8 (where he pines for Comum). On this occasion,
at least, country life is less attractive: he is not enjoying the work of
revising his speeches and, because he is in Tuscany, he cannot escape
the responsibilities of ownership and patronage — he was patron of
Tifernum-on-Tiber (IV. 1), and some at least of the 'rustici' who came
with 'libelli' may have been townsfolk requiring Pliny to act in disputes
for them. (It was normal that influential people, right up to the emperor
himself, should be lobbied by people with *libelli:* see, for instance,
Fergus Millar, 'Emperors at work', pp. 9ff.)

It would be easy to conclude from this letter that Pliny was a per-
functory manager of his estates. However, numerous letters show that
he took estate management very seriously: III. 19*; VIII. 2*; one to
his wife's grandfather (VI. 30*) showing how each was looking after the
other's properties; and a letter to Trajan (X. 8*). Pliny's wealth came
from his estates (annual income from his property at Tifernum alone
brought in 400,000 sesterces) and for this reason he could not afford
to neglect them.

Questions

1 For what purpose had Pliny 'taken refuge' in Tuscany?
2 Why did his plan go wrong?
3 Who are the 'rustici'? What do you imagine their complaints
   might be about?
4 What work of his own is Pliny doing? Why is he not enjoying
   it?
5 How is he sometimes breaking the monotony?
6 Explain the double meanings of 'rusticis' and 'urbana' (line 10).
7 Do you think Pliny is displaying an inconsiderate attitude towards his tenants?
8 In what way does this letter modify earlier impressions of
   Pliny?

IX. 27

The letter

An awkward incident during a recitation leads Pliny to reflect on the power and permanence of a historian's work.
See VIII. 16 for Paternus.

Notes

2 **historiae:** 'history-writing' not 'history' in its wider, modern sense.
3 **verissimum:** i.e. candid and forthright.
3-4 **partemque . . . reservaverat:** perhaps, in view of what follows, the writer had deliberately stopped at an interesting point.
5-7 **tantus audiendi . . . erubescunt:** 'so great is their shame at hearing what they have done, though they had no shame in doing what they blush to hear'. Supply 'eis' before 'quibus' and a second 'pudor' after 'nullus'.
9 **tanto magis quia non statim:** i.e. it will be read all the more because people have had to wait for it.

Exploration

It is just possible that the history being recited was Tacitus' *Histories*: the second half of this work was published about this time (Syme, *Tacitus,* I. 120). The power and dignity which Pliny attaches to the work being recited is consistent with his admiration for Tacitus; however, all this is conjecture.

What is certain is that the history recited on this occasion dealt with very recent events, and, to guess from the embarrassment it caused, events during the reign of Domitian. The 'amici cuiusdam' (line 4) are apparently the friends of someone — someone dead or even possibly still alive and prospering under Trajan— whom the historian was incriminating in some disgraceful behaviour, and they felt themselves also to be associated by implication.

Why were such people still at large? The answer is that after Domitian's death there had been no revenge taken on men who had compounded with his régime, and some of them had been able to continue in prosperous public careers.

Pliny was always deeply interested in literature, and had a particular regard for history, which he felt would bring immortality both to the writer and to the people described. In VII. 33* he tells Tacitus: 'I believe your histories will be immortal: a prophecy that will surely prove correct.' He ends the same letter: 'History ought not to depart from the truth, and truth is all that is required for honorable deeds.' He also thinks that history has a particular appeal to the public: 'Ora-

tory and poetry win small favour unless they reach the highest standard
of eloquence, but history cannot fail to give pleasure however it is presented.' (V. 8*).

The history presented at this *recitatio* was doubtless giving pleasure,
but not to everyone in the audience.

Questions

1 What qualities does Pliny say history-writing has?
2 What request was made at the recitation? By whom? What was
   their interest in the matter?
3 What does Pliny say was the inconsistency in their attitude?
   Do you agree?
4 Was the request granted?
5 What do you understand by 'sinebat fides' (lines 7-8)?
6 What does Pliny say the end result will be? Do you agree?
7 What should history set out to do?

IX. 33

The letter

Pliny tells the story of a tame dolphin at Hippo.
For Caninius Rufus, see II. 8.

Notes

2 **isto**: 'that ... of yours'.
   poetico ingenio: Pliny has already expressed an interest in Caninius'
   poetry writing (VIII. 4*). He now proposes this story of the dolphin as
   a suitable subject.
5 **vel ... scripturus**: 'even if you were intending to write ...' — the future
   participle is substituted for a conditional clause.
6 **Hipponensis colonia**: identified by Pliny the Elder as Hippo Diarrhytus
   (Zarytus) (*Natural Histories*, IX. 26), on the North African coast not far
   from Tunis.
7-9 **ex hoc ... stagno**: it was a salt-water lagoon connected to the sea by
   a tidal channel.
10 **etiam natandi**: the wording suggests that Pliny did not regard sea-
   bathing as a common pastime.
15 **praecedere**: historic or 'vivid' infinitive, the first of a series in this sentence, used for a fast, graphic narrative.
16 **trepidantem**: *sc.* 'eum' or 'puerum'.
21 **si quid est mari simile**: the humorous way of describing the eager curio-
   sity of the spectators: 'they gazed at the sea or at anything that looked
   like the sea'.
   ille: the boy who had already ridden the dolphin.

22 **delphinus rursus . . . puerum:** an elliptical sentence, again to make for
    fast narrative: 'the dolphin returned at the expected time, and again
    made for the boy'.
24 **variosque . . . expeditque:** 'and wove round and about in circles, this
    way and that'.
27 **praebentem:** here the word is used intransitively.
29 **nanti:** *sc.* 'delphino'.
37-8 **extrahi . . . siccatum . . . revolvi:** these three passives are used rather
    like the Greek middle voice: '. . . was accustomed to drag itself onto
    dry land, and, when it grew hot after drying itself on the sand, to roll
    back into the sea'.
39 **Octavium Avitum, legatum proconsulis:** nothing more is known of this
    man. The proconsuls of Africa and Asia were each allowed three *legati*
    (or assistants), men normally of praetorian rank.
    educto: *sc.* 'delphino'. It is Form C (dative).
40 **religione prava:** possibly Octavius thought the dolphin was a god in
    animal form, understandable if he was a devotee of Isis. For Egyptian
    beliefs on such matters, see Herodotus, *Histories,* II. Here, as in X. 96,
    Pliny shows prejudice against a religion which he does not understand.
41-2 **nec nisi post multos dies . . . :** 'and it was not until after many days that
    it appeared, listless and miserable . . . '
45 **novibus sumptibus:** the Roman provincial staff who arrived to see the
    spectacle expected to be accommodated and boarded at the expense of
    the locals.
47 **placuit:** 'it was resolved' — the word implies an official decision by the
    town council (cf. IV. 22, line 24).
    ad quod coibatur: noun clause subject of 'interfici'.

Exploration

It is not easy to say where the Younger Pliny came across this story.
Perhaps he used several sources: after all, he writes 'constat' (line 38).
He says he heard the story over dinner (lines 2-3). From whom? It may
have been from his uncle, but there are problems (see S-W, 514): the
uncle had been dead for a generation, if we accept that this letter is one
of Pliny's later ones; and the Latin (lines 1ff.) seems to suggest that he
has heard the story recently.

    Certainly the Elder Pliny gives the same story, in a shorter version,
in his *Natural Histories* (IX. 26); but there are differences of detail. For
example, the Elder Pliny says that it was the proconsul of Africa,
Flavianus, who poured oil on the dolphin and not his *legatus* (see letter,
lines 39-40). (This, incidentally, helps date the incident: Flavianus was
Tampius Flavianus, *legatus* of Pannonia in AD 69 and proconsul of
Africa early in Vespasian's principate [Tacitus *Histories,* II. 86; Syme,
Tacitus, II. 593].)

    It is unlikely that Pliny, with his devotion to literature and his *pietas*
towards his uncle, had not read his uncle's writings; but the divergence
of detail suggests that he came across the story elsewhere too. So how
is 'super cenam' (line 3) to be explained? Either by saying that one of
the occasions when he came across the story was a dinner-party; or that

the dinner-party is entirely fictitious — a literary device either to whet the appetite of the general reader or to persuade Caninius that Pliny has specially chosen this, from many stories, as the one most suitable for Caninius' pen.

The story itself, and so this letter, has no serious theme such as moral advice or the duties of patronage. It is straightforward narrative of the kind at which Pliny excels (see, for instance, the two Vesuvius letters, VI. 16 and VI. 20*). A reader cannot fail to note Pliny's economy: in just ten lines (5-14) he gives a full description of Hippo and its site, the habits of the inhabitants and the contests among the boys.

A teacher's very familiarity with this letter may hold in it the danger of taking the narrative too much for granted. Students must understand at every point what is going on in the story. In particular, the exact nature of the river estuary and the lagoon needs some working out. Such understanding of what is happening enhances the tale, whose charming blend of the fabulous and the everyday is the justification of this letter.

Questions

1 How does Pliny try to express the strangeness of his story?
2 Why is he writing about it to Caninius?
3 Where has he heard it?
4 Explain the words 'quid poetae cum fide' (line 4).
5 Describe the scene of the events.
6 Why were the boys in particular in the habit of going to the lagoon?
7 What game did they usually play?
8 Why did one boy swim out so far?
9 What did the dolphin do with this boy?
10 What did the townsfolk do when they heard about it?
11 What did the boy do the second time the dolphin appeared?
12 In what ways did the dolphin begin to show its tameness?
13 What other unusual thing did it do?
14 What did Octavius Avitus do? How does Pliny explain his action? Quote the words he uses. What do you think might have been in Avitus' mind?
15 How did the dolphin react to the attentions of Avitus?
16 Why did the whole thing begin to go sour? Who suffered?
17 Who decided to put a stop to it all? What word suggests the answer? Do you blame the decision?
18 How does Pliny suggest that Caninius should deal with the story? Do you feel the suggested style would be an improvement on Pliny's own account?
19 Do you believe the story? Does the behaviour of the people sound convincing? Are dolphins capable of doing such things?

IX.39

The letter

Pliny wishes to renovate the temple of Ceres on his estates, and writes
to enlist the help of Mustius.

For the various letters in this Selection on patronage, see I. 19.
Nothing is known of Mustius.

Notes

1-2 **in praediis:** presumably either at Tifernum-on-Tiber or at Comum.
Since the temple of Ceres was on his own land, Pliny did not have to
seek anyone else's permission to improve it. When he built the temple
at Tifernum-on-Tiber (dedicated in IV. 1), he needed to ask the town
council to find him a site (X. 8*).

3 **idibus Septembribus:** 13 September. Ceres was an ancient Italian corn-
goddess; and her September festival day referred to here helps point to
the connection of harvests and religious festivals going back to pre-
Christian times.

4 **multae res aguntur:** i.e. such as religious rites: Pliny is deliberately
vague.

5 **multa vota . . . :** 'many vows are undertaken, and many are discharged'.
This refers to the Roman (and Greek) custom of vowing to make some
gift or sacrifice to a god if he grants some favour.

6-7 **videor . . . facturus:** a colloquial usage: 'it seems to me that I will be
acting . . .'

8 **illam . . . has:** 'the former . . . the latter'.

10 **videbitur:** 'videri' is again used colloquially, this time to mean 'seem
good'.

13 **quantum ad porticus:** 'as far as the porticoes are concerned'.

20 **qui soles . . . :** 'for you are accustomed . . .'

Exploration

The advice of the soothsayers has presumably come as the result of a
bad harvest (or several) and Pliny decides to act on it immediately.
Mustius, who must be an architect (line 20), is asked to buy marble for
the floor and walls of the temple, and four columns for the frontage;
he is also asked to see to the making of a new statue of Ceres. Although
Pliny does not say so, these operations will necessitate a visit to the site
by Mustius.

The temple itself, the really urgent task ('ad usum deae', line 8), is to
be started at once. The portico, which is for the people, can apparently
be done later, and Pliny merely asks Mustius to draw up a plan for it.

The festival of Ceres in September was one of the most important to
country people, who on that day gave thanksgiving to the goddess for
their harvest, fulfilling the vows of sacrifice they had made long before

and making new vows for the following year. So on that day 'a great crowd of people assembles from the whole area' (line 4).

In view of such a great crowd, was not the temple rather small, with perhaps only four columns for the frontage (line 9)? Here students must remember that a temple was not built to accommodate all the people: a Roman festival took place in the open air, and the short culminating ceremony inside the temple involved the priests and acolytes only. It is for this reason that Pliny wishes to build the portico as a refuge for the crowd in case of severe weather. He gives no details of its construction, but it might have to be larger than the temple itself.

It is not clear where Pliny is writing from: possibly he is in Rome and has received the soothsayers' advice at second hand — perhaps from one of his family; Mustius is probably on the spot, since he is asked to buy the materials and draw up the plan.

Questions

1 What has Pliny been advised to do? By whom?
2 Why is it a good idea anyway?
3 For what favours would vows be given to Ceres?
4 To what festival in the Christian Church did the festival which is mentioned here correspond?
5 What additional building is Pliny going to provide? Where will it have to go, and why?
6 What three things does he ask Mustius to do?
7 Describe the site of the temple.
8 What is the profession of Mustius?
9 What is Pliny's motive for doing all this?
10 Do you find any evidence for Pliny's religious beliefs in the other letters you have read? Do you think he had any?

BIOGRAPHICAL NOTE ON THE EMPEROR TRAJAN

Trajan was born *circa* AD 53. His family came from Italica in Spain but was probably of Italian ancestry, from Umbria (Syme, *Tacitus*, II, Appendix 81, 785-6). Trajan's father was M. Ulpius Traianus (*consul suffectus* ?AD 70). He commanded the X *Fretensis* legion under Vespasian in the Jewish War and subsequently may have played an important part in Vespasian's policy in the east (G. W. Bowersock, 'Syria under Vespasian', *JRS* LXIII, (1973), 133-40). He was governor of Syria, Asia and his native Hispania Baetica.

The future Emperor Trajan was, according to Pliny (*Panegyric*, 15), a military tribune for ten years (though that may be an exaggeration), part of that being with the Syrian army. Subsequently he became quaestor (*circa* AD 78), praetor, *legatus* of the legion in Hispania Tarraconensis, helping to put down the revolt of Saturninus (AD 89), and then consul in AD 91. There is an inconvenient silence in the *Panegyric* about Trajan's career from AD 89 till the death of Domitian (AD 96). Nerva (Emperor AD 96-8) made him his heir in AD 97 and together they became consuls for the start of AD 98. Nerva died on 25 January AD 98, and Trajan received news of his death when he was in command of Upper Germany.

Trajan came to Rome early in AD 99 and was consul for the third time in AD 100 (with Julius Frontinus). In AD 101 he invaded Dacia and defeated Decebalus the following year. Decebalus kept his kingdom and in AD 105 renewed hostilities. This time Trajan annexed Dacia as an imperial consular province (AD 106), Decebalus having committed suicide. In October AD 113 Trajan set off on his eastern campaign. He annexed Armenia and then went on to attack Parthia. Though he captured the Parthian capital, he had to deal with rebellions and, turning towards home, he died in Cilicia in AD 117.

On the day after Trajan's death it was announced that the dead Emperor had named another Spaniard, Hadrian, as his heir. Hadrian had long been close to Trajan and his wife Pompeia Plotina, who had had no children.

Trajan was a great builder. At Rome he built new baths, the Aqua Traiana, the Naumachia and a vast new forum. The treasure of the Dacian Wars no doubt financed (or helped finance) this. Trajan's Column, close by the Roman Forum, illustrates on its spiral reliefs the Emperor's Dacian Wars.

The impression of Trajan gained from his rescripts in Book X is in line with the accepted tradition of him as a capable, concerned and humane Emperor.

X.31

The letter

Pliny has discovered that a number of convicts are now respectably acting as public slaves: torn between humanity and concern for the law, he consults Trajan.

Notes

1 **domine:** some emperors had shrunk from being called 'dominus'. Augustus 'domini appellationem ut maledictum et obprobrium semper exhorruit' (Suetonius, *Divus Augustus,* 53); and Tiberius would also have none of it. But it was 'a common form of polite address between inferiors and superiors of free birth' (S-W, 557), and it began to appear in official papers. Domitian apparently demanded its use by equestrian procurators in documents to him; and *ILS* 5795 quotes two letters of the mid second century AD (for one of which see, X. 37/8, 'Exploration'), where the *legatus* of the African army is addressed as 'domine', once by the city of Saldae and (presumably) a procurator, and the other time by a procurator. It is used by a junior to a senior senator in the will of Dasumius. From such evidence as is given above, it would seem a natural and by no means obsequious form of address from Pliny to Trajan. For a fuller discussion see S-W, 557-8.

2 **ius mihi dederis referendi ad te:** this suggests that Trajan had made a point of telling Pliny, perhaps in his *mandata* (see X. 96), to consult him without hesitation whenever he wished. Whether this was normally said to a governor-elect or only to Pliny in virtue of his special appointment is a matter of speculation: so too is the question of whether Pliny's letters to Trajan (sixty in just about two years, i.e. about one a fortnight) were more frequent than those of other governors. For a discussion, see S-W, 557-8.

3 **Nicomediae et Nicaeae:** Nicomedia, the capital city of the province of Bithynia and Pontus, had been founded in *circa* 264 BC by Nicomedes I. It had fertile lands and a good harbour and was on the main road from the Balkan provinces to the eastern frontier. For this city's abortive attempts to build an aqueduct, see X. 37/8.

   Nicaea had been founded earlier still, at the end of the 4th century BC and had soon become part of the kingdom of Bithynia. There was persistent rivalry between the two cities for the leading position. For the cities of Bithynia and Pontus, see A. H. M. Jones, *The cities of the eastern Roman provinces,* chapter VI.

4-5 **quidam . . . poenarum:** the people condemned had been provincial subjects (*peregrini*), not Roman citizens. *Damnatio in ludum* was condemnation to become a gladiator (see Michael Grant, *Gladiators,* Penguin 1971, 29-30, where this letter is discussed). *Damnatio in opus* meant condemnation to work in the mines. Condemned persons lost their civic status, and the people in this letter had not regained it by acquiring posts as public slaves; but they had managed to escape from savagely punitive sentences, condemnation *in ludum* meaning probable death, and condemnation *in opus* being a life sentence. These punishments were a creation of the emperors and are first attested here and in X. 58*.

  6 **publici servi:** public slaves were used for routine municipal work in Roman towns. This letter shows that some at least were paid; and X. 19* that public slaves were employed as prison warders.

12 **pasci:** this suggests just keeping them in prison – not a prescribed punishment but apparently sometimes used by governors (Ulpian, *Digest*, cited by John Crook, *Law and life of Rome*, 274).

18 **nulla monumenta:** *sc.* 'proferebantur'.

19-20 **iussu proconsulum legatorumve:** Bithynia being a public province, the governor was normally a proconsul with one *legatus* (assistant). Pliny kept the normal Bithynian quota of one *legatus* (Servilius Pudens, for whom X. 25* and S-W, 594-5), although he himself bore the special title of *legatus pro praetore* (see Appendix no. 1 and Introduction, pp. 14-16).

Exploration

Pliny could have settled this matter without reference to the Emperor simply by sending back the convicts to their punishment. However, he hestitates partly from humane feelings towards these men who are now, he says, 'frugaliter modesteque viventes' (lines 9-10), partly with the view that they were in no further need of punishment and perhaps partly because he felt an act of clemency might convince the provincials of his good intentions. On the other hand, he knew that to ignore the matter altogether would undermine his authority as governor of a province much in need of correction.

With these feelings he writes to Trajan requesting what is in effect an act of mercy which he himself has not the authority to bestow. Pliny is writing here in the considered belief that the case needs special attention from the Emperor, as with X. 33*, where he asks for the edict forbidding *hetaeriae* in Bithynia (X. 96) to be waived so that a fire-brigade might be recruited.

Students may find it as difficult as Pliny did to say what was the right course of action and it is certainly worth discussing what Trajan's reply ought to be in order both to preserve the rule of law and to be showing humanity.

The virtues of the Latin being read aloud are well demonstrated in this letter, especially in the sentence beginning 'nam et reddere poenae' (line 8), which becomes much clearer when the pauses are observed and some emphasis put on the words 'et . . . nimis severum, et . . . non satis honestum'.

Questions

  1 What does Pliny say the Emperor has given him permission to do?

  2 What problem has Pliny encountered?

  3 What four courses of action does he say are open to him? What are the objections to each of these?

4 What action has he taken in the meanwhile?
5 What enquiry has he made? Has it been successful?
6 What story did the men in question tell? Did Pliny find it plausible?
7 Which would have been the easiest course for Pliny to follow so that he kept within the law and avoided consulting the Emperor? Why has he not taken that course?
8 What do you think he ought to have done?

X.32

The letter

Trajan reminds Pliny that he has been sent to correct this kind of abuse; however, he is prepared to show a certain amount of clemency.

Notes

4 **ea:** 'from it' i.e. from the punishment.
   sine auctore: Trajan contradicts the argument of Pliny's last sentence.
8 **si qui vetustiores ... et senes:** the meaning is unclear. In the gloss we have taken 'vetustiores' as pretty well synonymous with 'senes'; Sherwin-White however suggests that 'vetustiores' means convicts 'of longer standing' (S-W, 606). If this is so, then Trajan is showing clemency not only to old men but to anyone condemned more than ten years before.
10 **solent:** 'it is usual for them ... '

Exploration

Trajan's reply is a compromise: he meets Pliny some way by allowing that the older, and perhaps the more long-standing, criminals who are now public servants be spared a return to their former sentences in the mines, but gives them more unpleasant tasks than they have already; the rest are to go back to work out their sentences. Thus some humanity is shown and at the same time the central government demonstrates that the law must be respected in Bithynia.

Questions

1 Of what does Trajan remind Pliny? Comment on the tone of his words.
2 What is his reaction to the news Pliny has given?
3 What is his decision? Do you think it is a cruel one? Or is it necessary?
4 Do you think Pliny might have been pleased with this reply? Would it satisfy his self-respect?

5 Trajan says elsewhere: 'I believe the people of the province will realise that I have their interests at heart' (X. 18*). Does his decision here accord with such a claim?

6 Might his decision involve the return to punishment of men who have genuinely been pardoned? Does Pliny think so?

7 What do you think was the purpose of this kind of punishment? What do you think is the purpose of any punishment?

X. 37, 38

The letters

The people of Nicomedia have spent great sums on an unfinished aqueduct. Pliny proposes finishing it and asks Trajan to send out an engineer.

Trajan is more concerned about what has happened to the money which has been spent, and directs Pliny to look into this. He approves of Pliny's scheme but fails to mention an engineer.

The other letters in this Selection concerning schemes of water-engineering are X. 41/2; X. 61; X. 98/9.

Notes: X. 37

1 **domine:** see note on X. 31, line 1.

1-3 **Nicomedenses impenderunt . . . :** comparative figures for other buildings may be informative: in Hadrian's reign 7,000,000 drachmas were spent on the aqueduct of Troas (i.e. nearly ten times the 3,318,000 sesterces already spent at Nicomedia) (S-W, 614). Pliny found that more than 10,000,000 sesterces had been spent on the unfinished theatre at Nicaea (X. 39*). The 200,000 sesterces (line 3) must be taken as the first instalment of intended payments towards a second attempt at the aqueduct.

It may come as a surprise that the leading city of Bithynia lacked a proper water-supply. However, eastern cities were far behind western cities in such amenities. Over a hundred years earlier the Pont du Gard had been built as part of the water-supply to Nemausus (Nîmes) in Gallia Narbonensis; yet in Pliny's time Sinope, as well as Nicomedia, had no aqueduct (X. 90*), and Alexandria Troas relied until Hadrian's principate on wells and cisterns. See S-W, 613-14 for a full discussion of water-supplies to towns in the east. For the city of Nicomedia, see notes on X. 31.

2 $\overline{\text{XXX}|\text{CCCXVIII}}$: students may ask about the horizontal and vertical bars used in writing this sum of money. $|\quad|$ means multiply by a hundred thousand; $\overline{\quad}$ by a thousand. So

$$\overline{\text{XXX}|} \quad = \quad 30 \times 100,000 = 3,000,000$$
$$\overline{\text{CCCXVIII}} = 318 \times \quad 1,000 = \quad \underline{318,000}$$
$$3,318,000$$

(see Hardy's commentary on this letter)

adhuc: qualifies the participle 'imperfectus'.

destructus: the stone had apparently been removed for future use with the second aqueduct ('lapide quadrato . . . detractus est': lines '9-10).

7-8 **arcuato opere, ne . . . perveniat:** the water was evidently to come down in an open channel, and so, not being under pressure, it would have to be kept at a high level until it reached the city.

Notes: X. 38

2 **ea:** goes with 'diligentia'.

5-6 **ne . . . reliquerint:** 'ne' means 'in case', and the emphatic clause, rather oddly, is 'dum inter se gratificantur'. The sense is 'in case it is for mutual pocket-lining that they have been beginning and abandoning aqueducts'.

7 **itaque:** 'and so': one of the rare examples of this word not coming first in its clause.

Exploration

In addition to the normal duties of a governor, Pliny had a special responsibility to investigate the accounts of the various cities; and it may be conjectured that, while he was investigating the expenditure of Nicomedia, he came across the two sums of money mentioned (3,318,000 and 200,000 sesterces) and so discovered that the Nicomedians had had two unsuccessful attempts at building themselves an aqueduct. Or did he see the unfinished arches as he travelled into Nicomedia, which led him to make inquiries?

At any rate, his interest was at once aroused, water-engineering being a subject close to his heart. He had been a friend of Julius Frontinus, who had written a book on aqueducts; and he himself had been from AD 104 to 107 *curator alvei Tiberis et riparum et cloacarum urbis* (Appendix no. 1).

Perhaps in writing to Trajan with this particular request, Pliny hoped to be sent an architect or water-engineer who had served on his staff in Rome when he had been *curator alvei Tiberis*. However, Trajan ignores Pliny's request altogether. This is at first sight surprising, in view of his agreement that Nicomedia must be given a water-supply (X. 38, lines 1-2). But the reason for Trajan's apparent lack of cooperation may be found in X. 18*, in *CLC* Unit IV 'Bithynia', where in answer to Pliny's request for a surveyor, he says that he has scarcely sufficient for the work in Rome and around. And Trajan was indeed a great builder (for his building activities at this time, S-W, 585-6). In X. 40* Trajan actually refuses to send an architect out from Rome — Pliny can find one himself. In X. 42 he tells Pliny he can apply to Calpurnius Macer, the governor of Lower Moesia, for a surveyor/engineer (*librator*) and ádds

that he will send out someone expert in this kind of work. In X. 62*
he omits to mention such an expert again, in spite of Pliny's reminder
(X. 61).

In not sending out men from Rome, Trajan may have been following
standard practice. Some forty years after this time a local veteran soldier, who was a *librator* of some competence, was employed on the construction of an aqueduct for Saldae in the North African province of
Mauretania Caesariensis (*ILS* 5795).

Part of this inscription is a short letter which the city of Saldae, and
presumably the procurator, sent to the *legatus* of the African army requesting the services of this particular *librator:* 'et Salditani civitas
splendissima et ego cum Salditanis rogamus te, domine, uti Nonium
Datum veteranum legionis III Augustae libratorem horteris veniat
Saldas, ut quod relicum est ex opere eius perficiat. ' Nonius Datus'
journey to Saldae proved hazardous: he fell among thieves, who strip-
ped him and wounded him. But, as he himself says, 'evasi cum meis',
making it to Saldae and finishing their aqueduct for them.

Students could translate this little letter as part of a discussion of
water-engineering. Others of Pliny's letters in Book X dealing with
public works in Bithynia could also be read; and, where possible, use
should be made of slides such as those in the Units of *CLC* or photographs from books such as Michael Grant, *The world of Rome,* or Sir
Mortimer Wheeler, *Roman art and architecture,* showing great aqueducts
like the Pont du Gard or the aqueduct at Segovia, as well as the *thermae*
at places like Pompeii, Herculaneum and Aquae Sulis and the bath-
house complexes at forts along Hadrian's Wall. (At Lugdunum the
Romans employed a syphon-system rather than a conventional aqueduct.) It has been remarked of the Pont du Gard that in the presence of
such engineering we reach the innermost core of the Roman mind.
Starting with the actual monuments, discussion could branch out from
Roman engineering skill to the values of the Romans as against those
of other peoples. Why, when it came to water-engineering, had the
eastern cities lagged so far behind those of the western Roman empire,
especially since Greek architects were often employed?

If Pliny is more concerned with building an aqueduct, Trajan in his
rescript turns his attention to the sums of money already spent on the
project. He tells Pliny to investigate this expenditure, particularly for
signs of corruption. Here is another possible line of discussion. Corruption of the kind Trajan feared does occur, though less widely tolerated,
in the twentieth century, and comparisons can be drawn — contractors
and architects bribing public officials to get them contracts; inflated
estimates of costs in order to gain maximum profit; the use of shoddy
materials; and, in Roman times, fraudulent mis-statements of the actual
size of buildings and the amount of material used. These were the kind
of abuses that Trajan was determined to stamp out.

Questions

1 What do the Nicomedians want?
2 How many attempts have they so far made to get it?
3 Why does Pliny say an arched structure is necessary?
4 What has Pliny done personally to help?
5 What are the three different sources of material which Pliny
   proposes for the new structure?
6 What does he imply is the reason for the earlier failures?
7 In what two respects does he recommend his project to
   Trajan?
8 What is the purpose of Pliny's letter?
9 Does Trajan make any helpful suggestions as to the building of
   the aqueduct? Or is he more concerned with another aspect of
   the question?
10 What does he suspect is the reason for the earlier failures?
11 Is there any request in Pliny's letter which he ignores? Why?
12 Does any phrase suggest Trajan dictated this letter personally,
   rather than leaving the reply to his secretary?

X. 41, 42, 61

The letters

Pliny outlines a scheme to dig a canal from Lake Sophon to the head of
the Gulf of Izmid (X. 41).

Trajan in his rescript (X. 42) feels Pliny's scheme needs further investigation before he is prepared to commit himself.

Pliny writes again (X. 61), having now investigated the technical
problems of the canal scheme; and this time he is authorised by Trajan
to go ahead.

Notes: X. 41

1-4 **intuenti mihi . . . habitura:** this rather turgid first sentence requires careful reading aloud, emphasising the pause after 'magnitudinem'. Note
also that 'habitura' ('likely to possess') goes with 'opera'. 'When I consider the greatness of your position and of your character, it seems
most appropriate that projects should be brought to your attention
which are worthy of your immortal fame as well as your present re-
nown, and which are likely to have as much beauty as utility.'

4-5 **est in Nicomedensium finibus amplissimus lacus:** for the city of
Nicomedia, see X. 31. Lake Sophon (Lacus Sunonensis) was, at its
western end, about eighteen miles east of Nicomedia. Its natural drainage was by a long river which flowed north-east into the Black Sea; but
Pliny wants to build a canal to link the lake with the Gulf of Izmid.

This would provide a waterway from the interior to Nicomedia, and do away with the inconvenient necessity of transporting goods partly overland (lines 5ff.). The letter eloquently shows the superiority of water-transport to land-transport in Roman times, even where there were good roads.

8 **hoc opus multas manus poscit:** there is a hiatus in the text here, in which Pliny has apparently suggested connecting the lake to the sea via the estuary at Nicomedia. Students may wish to supply the missing words: 'It seems possible that a canal might be dug from the lake to the estuary: this project would demand a great deal of labour.'

10 **certaque spes:** 'and the expectation is certain that ...'

11-12 **libratorem vel architectum:** see notes on X. 42.

13-14 **quadraginta cubitis:** Form E (ablative) of measure of difference.

15 **fossam a rege percussam:** this early attempt at a canal belonged to the days when Bithynia was still an independent kingdom (i.e. before 74 BC), but Pliny does not name the king.

17 **flumini:** the Sangarius estuary, flowing into the Gulf of Izmid at Nicomedia.

18-19 **intercepto rege ... effectu:** these Form Es (ablatives) absolute are causal: 'whether it was because the king was interrupted by death or because he despaired of completing the task'.

19 **feres enim me:** 'for you will tolerate my being ...'

21 **reges:** note the rhetorical plural.

Notes: X. 42

1-2 **potest nos ... velimus:** 'this lake of yours may possibly interest me so that I might wish to join it to the sea'.

2-3 **ne si emissus ... certe:** this clause may be put to the end of the sentence for translation purposes. It is a species of 'cautionary' clause and 'ne' may be translated as 'in case'.

3 **certe:** used in the apodosis of the condition, to strengthen the assertion. The meaning is almost 'as a consequence'.

4 **Calpurnio Macro:** Calpurnius Macer, also mentioned in X. 61/2 and X. 77, is known from documents to have been governor of Lower Moesia in AD 112-13. 'The reference to Macer provides the only direct evidence for the date of Pliny's mission' (S-W, 625).

4-5 **petere libratorem ... mittam:** on several occasions Pliny writes from Bithynia to ask Trajan for technical experts. For these requests and for Trajan's reaction to them, see X. 37/8, 'Exploration'.

Notes: X. 61

1 **domine:** see note on X. 31, line 1.

4-6 **potest enim ... dirimi:** 'For the lake can be brought right up to the river by a channel — not, however, let out into the river but both controlled and kept apart from it by leaving a sort of gap in-between.' Pliny visualises the canal coming to a dead end just near to the river — a very dubious scheme indeed and one which he does not mention again, partly because in any case he does not think it necessary.

5 **pariter:** here the word means little more than 'also'.

6 **viduetur:** *sc.* 'lacus' as subject.

8-9 **advecta fossa onera:** 'fossa' is Form E (ablative), while 'advecta' and 'onera' go together.

Exploration

It is important that students should have a clear picture of what Pliny is proposing (X. 41). The canal will have to run westwards from this inland lake for eighteen miles to join the estuary near Nicomedia. Trajan is afraid that the canal may dry up the lake completely, and so Pliny in the later letter (X. 61) suggests three possible schemes:

1 The canal could be dug as far as the estuary but not actually join it.
2 He could dam up the lake's natural drainage, the river which flowed out on its east side towards the Black Sea, and this would compensate for the loss of water via the canal.
3 A system of sluices could be used on the canal, and thus no water would be lost.

Trajan's tone in his rescript (X. 42) is rather cool, in contrast to Pliny's enthusiasm. Does he feel that Pliny is in danger of embarking on a grandiose scheme with less than his usual caution? Certainly the possibility which he mentions would be a disaster for the Bithynians: Pliny has explained the ease of water-transport compared to land-transport; and so, if the scheme were to result in the disappearance of the lake, the situation would be far worse than to begin with.

However, Pliny's technical investigations (X. 61) obviously impress the Emperor, who finally gives his consent to the scheme in a rescript of not untypical brevity (X. 62*).

These letters suggest a discussion of the Romans' love of water-engineering on the lines suggested in the 'Exploration' section of X. 37/8.

Questions

1 Distinguish the meanings of 'aeternitate' and 'gloria' (X. 41, line 3).
2 What is the lake already used for? What is the advantage of using it?
3 What is the frustrating part of the situation?
4 What improvement does Pliny suggest?
5 Where will the labour-force come from?
6 Why will the inhabitants be so keen on the project?
7 What expert does Pliny ask for? What will he have to find out? Does Pliny have any idea already of what his answer will be?
8 What unfinished project has Pliny discovered? To whom does he attribute it? Is he sure that it was for the same purpose as the one he is now suggesting? What two possible reasons does he give for its unfinished state?

9 In what terms does he tempt Trajan to undertake the project?
10 Do you know of any other building project of comparable size which the Romans undertook?

11 How does Trajan react to Pliny's scheme?
12 What possible disastrous result does he foresee? Would it be a blow to his prestige if it occurred? How would it permanently affect the Bithynians?

13 How in X. 61 does Pliny suggest ensuring against such a disastrous outcome? Do you think any of the suggestions would work (either the one in the Latin or those in the summary)?

X. 77, 78

The letters

Pliny asks Trajan to provide for Juliopolis a unit of soldiers such as he has allocated to Byzantium. Trajan refuses: he does not wish to set a precedent and instead advises Pliny on how to enforce the law.

Notes: X. 77

1 **domine:** see note on X. 31, line 1.
1-2 **Calpurnio Macro:** for this man, see X. 42. Lower Moesia where he was governor, was the nearest province to Bithynia with legionary troops.
2 **legionarium centurionem:** as these two letters show, this centurion had been sent to Byzantium to help preserve public order. We do not know how many soldiers he was given for this mission, one of the reasons for which may have been Trajan's Dacian campaigns and a consequent increase of traffic through an already busy city.
2-3 **Byzantium:** on the western side of the Bosphorus, but by Pliny's time part of the province of Bithynia, in which it stayed even when Trajan made Thrace a province of equal status with Bithynia.
3 **Iuliopolitanis:** Juliopolis was the frontier town on the only road into Bithynia from central Asia Minor. It had been a village until Cleon, a colourful rogue of a brigand, made it his headquarters, enlarged it and renamed it Juliopolis to commemorate Julius Caesar (Strabo, XII. 574-5).
5-6 **onera maxima sustinet tantoque graviores iniurias . . . :** Hardy, in his commentary, p. 188, believed that the 'onera' referred to were imperial customs dues levied extortionately by *publicani* away from the watchful eye of the Roman government. Certainly in view of its location there was likely to have been a customs point at Juliopolis. But

soldiers did not collect customs dues, and Trajan's rescript makes it
clear that soldiers are among those causing trouble to the Juliopolitans.
Although numbers of travellers passed through Juliopolis (X. 77, lines
8-9), the people who are being such a burden to the town are apparently soldiers or those on official government business ('contra disciplinam meam': X. 78, line 9 implies that the offenders were in the
Emperor's service).

Even the imperial courier service (*cursus publicus* or *vehiculatio*),
being paid for by local communities, was a constant source of grievance
(see S-W, 627-8 and CLC *Handbook* to Unit IV, pp. 63f.). But Roman
officials no doubt often compelled hospitality where it was not offi-
cially provided, and this is probably what was happening at Juliopolis.
Compare IX. 33, lines 43-5 where Pliny in his account of the dolphin
at Hippo speaks of the expense to the town incurred as a result of
visiting dignitaries; this is also mentioned by the Elder Pliny in his
version of the story, 'iniuriae potestatum in hospitales ad visendum
venientium' (*Natural Histories,* IX. 26).

5-6 **tantoque ... quanto:** it is impossible to translate this correlative construction literally, the difficulty being that Latin idiom throws both
adjectives into the comparative: 'because it is weak, it suffers injustices
all the more serious'.

6-7 **quidquid autem ... provinciae proderit:** it is difficult to get at the
precise point Pliny is trying to make here. In IV. 22 he speaks of evils
spreading 'a capite' (lines 27-8) to the whole body; and perhaps he felt
that, with a severe check kept on officials as they entered the province
at Juliopolis, such people would be more likely to be on their best
behaviour for the rest of their journey through Bithynia.

Notes: X. 78

2-3 **secundum consuetudinem ... :** as a preliminary to his refusal, Trajan
says that Byzantium has had such detachments before. Perhaps previous
emperors had sent them when military campaigns in the east had increased traffic through the city.

3-4 **honoribus eius ... consulendum habuerimus:** 'I thought I ought to support its magistrates ... ': 'honoribus' is an abstract noun substituted
for a personal one.

6 **plures:** *sc.* 'civitates'.
eo: 'on that account', taking up 'quanto', whose clause it would more
easily follow.

7-8 **fiduciam eam ... iniuriis:** 'I have sufficient trust in your conscientious-
ness to believe that you will act in every possible way to see that they
are not exposed to injustices.'

9 **disciplinam meam:** S-W, 668, for all service under the emperor being
militia even for civilians.

10-12 **si plus ... mihi scribes:** there was a limit to Pliny's *imperium* in
Bithynia: he had no capital jurisdiction over Roman citizens (see, for
instance, X. 96, lines 15-16), nor the *ius gladii* necessary for control of
legionary troops, the soldiers stationed in Bithynia being auxiliaries.
Trajan tells Pliny to refer legionary soldiers to their commanders,
'legatis' (line 11) being *legati legionum*. Those going up to the capital
were presumably not soldiers ('milites' belonging to the first 'aut'
clause) but men on government business (e.g. procurators).

Exploration

It is noticeable that Trajan's rescript is longer than Pliny's letter. Pliny adopts a familiar stance at the start: obviously he hopes that a compliment will help his case with Trajan: 'providentissime, domine, fecisti . . .' (lines 1-3). But after this first sentence he wastes no words in explaining the need for Juliopolis to have a military contingent such as Trajan has given Byzantium. He may have been so brief, just because he thought Trajan could not justifiably refuse this request.

Sherwin-White has commented interestingly on the tone of Trajan's rescript: the official, civil-service character of the first half, with the bureaucrat's horror of saddling himself with a precedent, 'followed by the crisp intervention of the Princeps on a point that interests him' (S-W, 667-8). It is surprising that Trajan has written at such comparative length — this rescript is longer than X. 97 which answers Pliny's voluminous letter about the Christians (X. 96). Trajan deals dexterously with Pliny's reasonable request, perhaps a little too dexterously. Pliny has pleaded that the smallness of Juliopolis makes it particularly vulnerable to abuses (lines 4-6); but Trajan turns this as a weapon against him — where will be the end, if even small towns are given contingents? And one may suspect that Trajan has deliberately parried 'onera' (X. 77, line 5) with 'onerabimus nos' (X. 78, line 5) — Juliopolis may have its burdens, but it will burden us with a precedent to give Juliopolis soldiers.

These neat semantics may conceal an unease in Trajan. Book X demonstrates beyond contradiction that both the Emperor and Pliny had the interests of the provincials at heart. It would no doubt help the people of Juliopolis to have a military contingent, but there were not enough soldiers for all these extra duties. (Trajan had already fought his successful Dacian campaigns but, because of casualties, units must have been below full strength. Meanwhile there would soon be the Parthian campaign with all its call on soldiers.)

Instead of soldiers, there are Trajan's precise instructions to Pliny (X. 78, lines 8 to end). These are a little too detailed, suggesting a real weakness. Twice before Trajan has stressed to Pliny that the troops within Bithynia should not be taken away from normal service (X. 20* in *CLC* Unit IV 'Bithynia'; X. 22*). So if no soldiers either from inside or from outside are to be sent to Juliopolis, who is to see that the existing abuses are checked? Certainly not Pliny himself: among other things Juliopolis is in a remote part of the province. One presumes that the town magistrates and private individuals must be the ones to take the initiative, thankful that they live in good times with a scrupulous man like Pliny as their governor.

It was the same everywhere. Only a small staff of Roman officials assisted the governor in each province; and if initiative was to be taken, it would not be by the government but by individuals and individual communities. (See Sherwin-White in *Roman civilisation*, ed. J. P. V. D. Balsdon, 88-9 for a discussion of this exchange of letters.)

Questions

1 On what does Pliny compliment the Emperor?
2 To what request is this a preface?
3 What are the arguments for this request?
4 What is Trajan's reply to the request?
5 Why does he say he has sent troops to Byzantium?
6 Is Trajan giving Pliny more work?
7 Who is to be responsible for law and order at Juliopolis?
8 Just how precise are Trajan's instructions?
9 Do you think Pliny was satisfied with Trajan's rescript?
10 Does this pair of letters suggest a weakness in the Roman provincial system? Does Trajan's tone suggest unease?

X. 94, 95

The letters

Pliny asks Trajan to grant to his scholarly and deserving friend Suetonius the *ius trium liberorum*. Trajan agrees.

For other letters in Book X concerning benefits for individual people, see X. 106; also X. 87*.

Notes: X. 94

1 **Suetonium Tranquillum:** Suetonius is best remembered now for his pocket biographies of Julius Caesar and the first eleven emperors, Augustus to Domitian. These, an idiosyncratic mixture of history and scandal, make compulsive reading: they are also excellent as short character-studies. Suetonius was born into an equestrian family between AD 70 and 75. He held the posts of *a studiis* and *a bibliothecis*, before becoming Hadrian's *ab epistulis*. He fell into disgrace *circa* AD 121 along with the Prefect of the Praetorians, C. Septicius Clarus, and others (see II. 9, note on lines 15-16). For Suetonius' career, see the fragmentary inscription, Smallwood: (2) 281; Syme, *Tacitus*, II. Appendix 76, 778-81). Pliny and Suetonius had a common interest in literature. Pliny wrote four letters to him: I. 18*, III. 8*; V. 10*; IX. 34*. In I. 24* Pliny is trying to negotiate an advantageous price for a small property which Suetonius wants to buy. Perhaps Suetonius was on Pliny's staff in Bithynia: see note on line 4.

3 **domine:** see note on X. 31, line 1.
  in contubernium adsumpsi: in I. 24* Pliny calls Suetonius 'contubernalis meus'; and Romatius Firmus was Pliny's 'contubernalis' too (I. 19). Pliny wants to show Trajan that he and Suetonius are close friends.

4 **quanto nunc propius inspexi:** this suggests that Suetonius was on Pliny's staff in Bithynia, as indeed does the inclusion of this letter in Book X.

4-5 **ius trium liberorum:** this privilege, the benefits of which are not fully known, was largely brought into being by the *Lex Papia Poppaea* of AD 9. Augustus instituted it with a view to encouraging the upper classes of Italy to have larger families, granting its benefits to those with three children and penalising those who did not. The benefits concerned rights of inheritance, and also included preference in standing for magistracies, seniority in office and in the allocation of proconsulships, as well as special seats in the theatre. A parent entered a child on the public register, which was presumably proof of the size of a family (J. A. Crook, *Law and life of Rome*, 46-7). Originally control of the privilege had rested with the senate; but by the Flavian period it had passed to the emperor. Where a man had three children, the grant must have been a formality. However, the grant of this privilege to those who had less than three children was no formality, to judge from the evidence of Trajan himself ('quam parce haec beneficia tribuam': X. 95, line 1) and of Pliny (who says that the Emperor Nerva gave it 'parce et cum delectu': II. 13*). It is impossible to say how often it was given. Pliny, though childless, obtained it (X. 2*); and, no doubt sympathising with Suetonius' own lack of children, he petitions the Emperor for him. He also obtained the grant for Voconius Romanus (II. 13*). For the *ius trium liberorum*, see S-W, 558.

5-6 **iudicia amicorum promeretur:** for those who were childless there were restrictions on their rights of inheritance. It seems therefore that Pliny is giving as one valid reason for Suetonius needing the *ius trium liberorum* the fact that he could then inherit from friends. For this use of 'iudicia', see Suetonius *Divus Augustus*, 66: 'amicorum tamen suprema iudicia morosissime pensitavit'. The context is Augustus' attitude to legacies.

7-8 **impetrandumque . . . denegavit:** 'and he has to obtain from your benevolence — at my request — what the cruelty of fortune has denied him'.

Notes: X. 95

3-4 **apud amplissimum ordinem:** i.e. before the senate.

4 **suffecturum mihi:** Trajan must be referring to an upper limit which he did not intend to exceed: 'quam parce' and 'non excessisse' make this clear.

6 **ea condicione, qua adsuevi:** we do not know on what conditions Trajan granted the *ius trium liberorum*; but they can hardly have included a time-limit such as so notoriously mean an emperor as Galba imposed (Suetonius, *Galba*, 14).

6-7 **in commentarios meos:** these were the Emperor's official records. In X. 66* Trajan refers to the records of his imperial predecessors, when dealing with a letter from Pliny ('in commentariis eorum principum, qui ante me fuerunt').

Exploration

The way Pliny commends Suetonius to the Emperor repays attention. In the first four lines he praises him; starting with a general commendation, Pliny reveals that admiration has led him to become a close friend of Suetonius and that, having got to know him better, he likes him even more. Only after the eulogy does Pliny explain why Suetonius needs the *ius trium liberorum:* it would enable his friends to show their appreciation, and Suetonius has not had a fertile marriage and so must look to the Emperor's generosity. In similar circumstances today a person might be tempted to put Suetonius' needs first and his own recommendation second. But Pliny knew how the system worked; and without any particular egotism he realised that, if Trajan were to accede to his request for Suetonius, it must be because he (Pliny), a consular and special envoy of the Emperor, is recommending him. The Emperor did not hand out benefits on a humanitarian basis but because individuals could be seen to be worthy of them.

Trajan sends Pliny a friendly ('mi Secunde carissime') but appropriately businesslike reply which calls for little discussion.

Students wishing to discuss the Roman habit of helping people in their careers should read II. 9, II. 13*, III. 2*, IV. 4*, VI. 23, VII. 22*, X. 87*. These letters are similar to those written today as references for a person applying for a job. But perhaps in a society as hierarchical as the Rome of Pliny's day, recommendation had a particular importance.

Questions

1 What do you know about Suetonius?
2 What is Pliny requesting for him?
3 Who else had the same privilege although he had no children?
4 On what grounds does Pliny make the request? Do they seem fair?
5 What has the cruelty of fortune denied Suetonius? In what sense can Trajan supply it? Does Pliny think the privilege will be adequate compensation?
6 In what terms does Trajan address Pliny? Does it contrast with Pliny's mode of address to him?
7 What does Trajan say he has told the senate about the granting of this privilege?
8 Translate the word 'desiderio' used by Trajan (line 5).
9 What was Trajan's purpose in granting people the *ius trium liberorum*?

X. 96

The letter

For the first time, Pliny has had to deal with people denounced as being Christians. He is unsure as to whether he has been following the right procedure, and writes to Trajan for guidance, describing what has been done so far.

For full discussion on this letter in general, and on individual points, see Sherwin-White (pp. 691-710; 772-87).

Notes

1 **domine:** see note on X. 31, line 1.

3 **cognitionibus:** the *cognitio* was a trial conducted in person by the holder of *imperium*, and the verdict was his personal judgement. He might be a provincial governor or, in Rome, the city prefect (S-W, 694-5). Pliny's wording implies that he knew of such trials of Christians taking place before.

4 **quid et quatenus:** these together qualify both the verbs 'puniri' and 'quaeri': 'in what way and to what extent'.

6 **an quamlibet teneri:** 'or whether the young, however young they are ...'

9 **flagitia cohaerentia nomini:** the Romans were usually tolerant towards alien cults and religions (see notes on X. 97). But they ceased to be tolerant when the devotees were shown to be practising anti-social acts, *flagitia* or *scelera*. So they punished the Druids and Bacchanals for murdering people in secret rites; and Christians were rumoured to indulge in cannibalism — an obvious misunderstanding of the rite of Holy Communion.

7-9 **detur paenitentiae ... puniantur:** *sc.* 'utrum' at the beginning of both these indirect questions.

10 **qui ad me tamquam Christiani deferebantur:** these words indicate that Pliny had not gone out of his way to track down Christians, but that an informer, or informers, had denounced them to him. This process of informing (*delatio*) was normal in the Roman empire, for there was no Director of Public Prosecutions and no police force to initiate prosecution. In his rescript Trajan tells Pliny that he must proceed against Christians only as a result of *delatio* — there must be no witch-hunt, nor must he take any notice of anonymous pamphlets (*libelli*).

13-15 **perseverantes duci ... debere puniri:** it appears that Pliny had been sentencing Christians to immediate execution (this is to be understood from the word 'duci'), not because of any crime implied by their being Christians but because of their obstinacy in not bending to his authority. Pliny was quite within his rights here, as was the governor of Sardinia who in AD 69, wary of a protracted dispute over boundaries, threatened to punish the Galillenses 'quodsi in contumacia perseverassent' (*ILS* 5947). Such action by a provincial governor was regarded, apparently, as an acceptable way of enforcing respect for Roman rule. It is interesting, however, that Pliny was not obliged to proceed so severely against the Christians. His *imperium* gave him, as governor, the power of life and death over non-citizens in Bithynia, but the choice was left to him. He could have passed a lighter sentence, or even ignored the charges. In the Julio-Claudian period the governor of Achaea, Gallio, refused to proceed with the case when the Jews brought St Paul to court (Acts of the Apostles, 18).

16-17 **quos, quia cives Romani erant . . . remittendos:** by Trajan's time, governors of provinces automatically sent Roman citizens on serious charges for trial in Rome.

18-19 **mox ipso tractatu . . . inciderunt:** 'soon, as accusation has become widespread because of the very investigation — the usual result — more kinds of instance have come to my notice'. Apparently people began to use denunciation to satisfy grudges or for other personal motives.

20-5 **qui negabant esse . . . putavi:** 'Those who denied that they were, or had been, Christians, I thought should be sent away, since they called upon the gods with me beginning the formula, and with incense and wine made prayer to your image, which I had ordered to be brought in (to court) for this purpose, along with the statues of the gods; furthermore, they cursed Christ; and it is said that those who are really Christians cannot be compelled to do any of these things.'

22-3 **imagini tuae . . . supplicarent:** worship of the emperor was not compulsory: Pliny is simply using it as a test to discover Christians.

25-8 **alii ab indice nominati . . . dixerunt:** although Pliny does not say so, the implication of the whole letter is that these apostates are being held with their case undecided until Trajan has been consulted.

30-6 **adfirmabant autem . . . abnegarent:** 'However, they declared that the sum total of their guilt or error had been this: that they had been in the habit of meeting before dawn on a regular day, and reciting, in turn with each other, a hymn to Christ as if he were a god, and binding themselves with an oath, not to any crime but to refrain from committing theft, robbery or adultery, not to break a promise, and not to withold a loan when called upon to restore it.' These lines give a valuable picture of the pattern of Christian worship at that time, but one capable of several interpretations (see S-W, *passim*).

31-2 **stato die:** i.e. Sunday, the day after the Jewish Sabbath. The week as a division was unknown to the Romans, except those who were familiar with Jewish customs.

32 **ante lucem:** this does not necessarily imply an attempt at secrecy: people used to start their day before dawn. So Pliny the Elder went to Vespasian 'ante lucem' (III. 5*).

33 **invicem:** the Christian practice of chanting psalms and canticles antiphonally, a variation of earlier Jewish practice.
sacramento: to a Roman this word had the commonest connotation of a 'military oath' and, because to Trajan this might suggest some sort of secret military organisation, Pliny hastens to add 'non in scelus aliquod . . . ' It is very probable that the prisoners had used the word 'sacramentum' in its special Christian sense of 'sacrament', and that Pliny had not realised its esoteric use; but it is also possible that it is his own choice of word to describe what they narrated to him.

34-6 **ne furta ne latrocinia . . . :** this may possibly be taken as evidence that the Ten Commandments were read at the early liturgies of the Mass, though it could simply refer to the general tenor of the service.

35-6 **ne depositum adpellati abnegarent:** in an age when banks were not so widespread, the safekeeping of a friend was a way of having one's money looked after, and evidently fraudulent conversion of such deposits was so common that Christians bound themselves explicitly not to be guilty of it.

37-8 **rursusque coeundi . . . innoxium:** this can be taken as a social occasion, when the Christians met on a Sunday to break bread together; or it could be taken to describe a celebration of the Mass later in the day, perhaps in the evening. However, the service 'ante lucem' sounds more like the Mass because (1) it is given a more important mention; (2) it contains the word 'sacramentum'; (3) lines 38-9 suggest that the Christians gave up the later meeting of the day in obedience to Trajan's edict, which they would hardly do if it were the Mass.
promiscuum tamen et innoxium: i.e. not a meal of human flesh: cannibalism was one of the rumoured practices of Christians.

38-9 **post edictum meum . . . vetueram:** Pliny had issued an edict on his arrival in the province. From early times it had been the duty of a praetor at Rome to issue an annual edict on entering office. Rome's provinces were at first governed by praetors in office and it may be presumed that the practice of a governor to issue an edict in his province was derived from the praetor's edict at Rome. (For Cicero's edict for Cilicia, see for instance *Ad Familiares,* III. 8. 4.)
mandata: the confidential instructions issued by the emperor to governors-elect of his own imperial provinces and later (probably after Pliny's time) to governors-elect of senatorial provinces also. Pliny in any case received *mandata* by virtue of being the Emperor's special appointee. *Mandata* seem to have consisted partly of instructions applicable to all (or many of) the provinces and partly of special rules applicable to a governor's own province (so, here, Trajan had included in his *mandata* to Pliny the instruction that *hetaeriae* be suppressed in Bithynia and Pontus). For *mandata,* see Fergus Millar, 'The Emperor, the senate and the provinces', *passim;* S-W, 589ff. The banning of *hetaeriae* in Bithynia was a new ruling, which was later extended to other provinces (S-W, 608-9). Bithynia had been troubled by political unrest, which Trajan connects with clubs in general: 'sed meminerimus provinciam istam et praecipue eas civitates eius modi factionibus esse vexatas': X. 34*. Hence his ban: though he makes one exception by allowing mutual benefit societies in the free city of Amisus (X. 93*).

39-41 **quo magis necessarium . . . quaerere:** 'so I believed it all the more necessary to discover the truth, and to do so by torture, from two slave-girls who were said to be deaconesses'. Pliny is unable to believe from the prisoners' accounts that the Christians' activities can be so innocuous, and so he resorts to this barbarity. It has been suggested that the two girls were not really slave-girls at all, Pliny having misunderstood the Greek word διακόναι, deaconesses. However, it seems most unlikely that the meticulous and legally experienced Pliny would have taken the step of torturing people without the most careful verification of their status as slaves. (Students should note that the evidence of slaves was not legally admissible unless they were tortured.)

42 **superstitionem pravam:** a similar phrase is used for the action of the deputy governor who poured oil on the dolphin at Hippo (IX. 33): there, as here, Pliny shows no apparent interest in the religious motives.

47-8 **neque civitates tantum . . . pervagata est:** the Acts of the Apostles reveal how soon Christian missionary work had begun in the Asiatic provinces: St Paul and others had gone there to preach and convert in the Julio-Claudian period. So it is not surprising that by the early second century, Bithynia contained Christians in some numbers (see I Peter 1. 1).

49 **prope iam desolata:** the word 'prope' is used like 'fere': ' . . . which recently were almost deserted.'

51 **venire carnem victimarum:** Pliny regards the sale of sacrificial animals as a gauge of how many people are returning to the traditional pagan religion.

Exploration

Pliny apparently feels out of his depth and is seeking some guidance on policy from the Emperor. A first reading of this letter and its reply would seem to provide yet more evidence of Pliny's humanity and broadmindedness; but on closer inspection, it is Trajan who emerges as the more tolerant of the two men.

Why were Christians persecuted in this way, when the worship of Isis was encouraged, and even a religion as strict and unbending as that of the Jews was tolerated, at least by the good emperors? For a full discussion of this question, see Sherwin-White's fifth Appendix (S-W, 772-87).

It was Roman policy to tolerate and even encourage national cults: the local gods would be angry unless they were paid their due rites. But Christianity was not a national cult. Indeed, by its failure to associate itself with any specific country or place, Christianity disqualified itself from being regarded as a genuine religion at all, so deeply embedded in the Roman consciousness was the idea that the gods were operative in specific places. Rather, such a religion had the characteristics of magic or witchcraft.

Not only that, the Christians refused to recognise the validity of any other kind of worship, an incomprehensible attitude to a people brought up to polytheistic religions. Such strictness might be pardonable in the Jews who kept their ideas to themselves, but the Christians were actively trying to convert the other people in their communities. Pliny knows this when he calls it 'superstitionis istius contagio' (line 48).

Another, and important, consideration was the reputation which the Christians had acquired for themselves. Rumour attributed to them such practices as cannibalism, promiscuity and incest, and they generally were regarded as an anti-social organisation. Tacitus (*Annals*, XV. 44) takes it for granted that they deserved even the tortures which Nero had inflicted on them after the burning of Rome.

This letter is of interest in giving an outsider's account of early Christian practices — fragmentary and blurred by misunderstanding, but including

a. the custom of antiphonal singing (lines 32-3);
b. the *sacramentum*, which Pliny takes to mean just 'an oath';
c. the communal meal (line 37), which is not the Mass but simply a social gathering, which the Christians are willing to forgo to avoid breaking the law (lines 38-9).

Finally, there is the question of how Pliny's own attitude and character emerge in this letter:

1 He assumes that the Christians are wrong, but is unsure why. His trained legal mind suspects, rightly, that there is no actual law against being a Christian (lines 8-9: 'should a man be punished for just being a Christian, or only for committing the crimes associated with that name?'). He does know that the law against *hetaeriae* may apply, but Christians may not necessarily be breaking that law. He also makes the assertion that their 'pertinacia et inflexibilis obstinatio' (lines 14-15) ought to be punished. Of this he says, 'neque dubitabam' (lines 13-14), but such a statement is a little inconsistent with his hesitancy in other parts of the letter, and, it could be argued, it is a statement out of character with Pliny's *humanitas* as we discover it elsewhere. Is he aggressively trying to compensate for his indecision elsewhere in the letter?

2 Pliny uses the words 'amentia' (line 16) and 'superstitio prava et immodica' (line 42) about something he has apparently not even attempted to understand.

3 He clearly implies (lines 19-20) that he has acted upon the contents of anonymous pamphlets, and is indeed forbidden by the Emperor to go on doing so.

4 He gives a rather unconvincing assurance that the trouble is already being stopped (lines 49-54). Is he again trying hard to assure the Emperor of his competence?

5 He seems to be quite unsure how Trajan is going to react, and so wavers between representing himself now as firm, now as merciful. In the end, as the rescript shows, it is Trajan who is the more lenient of the two.

Questions

1 On what three questions does Pliny say he is in doubt?

2 Why does Pliny think that the professing Christians deserve what punishment he is giving them?

3 What has he been doing with the Christians who are Roman citizens? Why?

4 What has caused the sudden rash of accusations? Do you think all of them were genuine? What motive would anyone have for making a false accusation of this sort?

5 How has Pliny been testing the veracity of those who deny being Christians?

6 Does the ex-Christians' account explain the actual beliefs of Christianity? What similarities are there between their description and modern Christian practice?

7 Was there anything in the activities they described which was contrary to Roman law? What law (see line 39)? Were the Christians forming a *hetaeria*?

8 Why has Pliny thought it necessary to torture the two girls
   (see note on lines 39-41)?
9 In what words does Pliny summarise the Christian religion?
   Does he anywhere in the letter show that he has tried to
   understand it?
10 Why is Pliny consulting the Emperor at all?
11 Does he think Christianity is already beginning to die out, as a
   result of his enquiries? What evidence does he give for this?
   Is it convincing evidence or not?
12 Does Pliny strike you as a thoughtful man on matters of religion? Does he talk about it in any other letters you have read?
13 Is this letter a sympathetic one? Do you dislike any parts of it?

X.97

The letter

Broadly, Trajan approves of the way Pliny has dealt with the Christians.
He says that there cannot be a fixed procedure, but he does specify that
there must be no witch-hunt nor any use of anonymous pamphlets.

Notes

2 **Christiani:** predicative: 'as Christians'.
  delati fuerant: by the double pluperfect Trajan means that the Christians 'had been in a state of having been denounced', that is, had
  already been dealt with when Pliny wrote his letter.
5 **ita tamen:** 'but with this proviso'.
9-10 **pessimi exempli . . . nostri saeculi:** characteristic Form Ds (genitives).

Exploration

Whereas we find Pliny rather too involved in the situation and ready to
be intolerant, Trajan is calm and unexcited, and ready to let the Christians go their own way so long as they do not cause trouble. In taking
such an attitude, however, he is not formulating a new approach. The
Roman government had a good record of tolerance towards alien cults,
broken only now and then by sudden reversals of policy. Even the
Christians, whose religion gave such grounds for suspicion to Romans
(see 'Exploration' on X. 96), normally benefited from this tolerance.
  However, the general lack of persecution of the Christians was not
based solely on principle. Much more of a reason is the way the Romans
ran their empire. In the imperial period the central government in Rome

did not go out of its way to look for problems. Trajan was adopting the standard attitude of letting problems come to him rather than go out chasing them. Furthermore, the legal system of *delatio* encouraged this approach, enabling as it did any person who felt strongly that the law had been broken to lay information before his local magistrates or the governor without there being any notion that in doing so he was behaving objectionably.

Anonymous denunciation, on the other hand, was a different matter. It was regarded then, just as today, as one of the practices associated with a tyrannical government, and Trajan thinks it would reflect adversely on his benevolent principate ('nec nostri saeculi est', line 10).

So Trajan has made two prohibitions: he forbids Pliny to hunt down the Christians, and he forbids him to use anonymous information. Apart from these, however, does he really answer Pliny's questions (X. 96, lines 5-9) or give the advice, to await which Pliny has actually 'postponed the enquiry' (line 43)? Can we even accuse the Emperor of giving an inadequate reply to Pliny's consultation?

In fact his answer is to be found plainly enough in the first four lines. Pliny has been conducting the affair as he ought ('debuisti'), but Trajan is not prepared to lay down a fixed procedure for general application. He then makes his two prohibitions, thus modifying his general approval of Pliny's methods.

Questions

1 Is Trajan able to give a general rule for how to treat Christians?
2 What two things does he forbid Pliny to do?
3 Explain what Trajan means by the last sentence.
4 Who seems to show more leniency in this affair — Trajan or Pliny?
5 Do you think that this letter was dictated by the Emperor personally? On what grounds?
6 Do you consider this to be an adequate reply to Pliny's request for guidance?

X. 98, 99

The letters

The beautiful city of Amastris has a foul sewer which Pliny asks permission to cover over. Trajan grants permission

For other letters in the Selection dealing with water-engineering in Bithynia, see notes on X. 37/8.

Notes: X. 98

1 **Amastrianorum civitas:** Amastris had been founded about 300 BC by a Persian princess of the same name (who had been deserted by two husbands and had become the widow of the third). It was first in the territory of Heraclea, then part of the kingdom of Pontus, which was to come under Roman rule.
  domine: see note on X. 31, line 1.
2 **eandemque:** 'and also' — an idiomatic usage.
3 **cuius a latere:** 'on whose side'.
4 **nomine quidem flumen:** 'what is in name a river'.
  cloaca: for Pliny's specialist knowledge of drainage, see Introduction II: Pliny's career, p. 8.
4-6 **ac sicut turpis ... taeterrimo:** *sc.* 'est' with both these clauses.
7-8 **quod fiet si permiseris curantibus nobis:** a compact expression: 'this will be done, if you grant me permission, and I will see to it that ... '

Exploration

Pliny's short letter is carefully phrased: he talks of the city's architectural beauty, then passes on to one of its most beautiful features, a long boulevard. Then, spinning out the wording for maximum effect, he reveals as a shock that what is in name a river running beside this fine street is in reality a noisome sewer, horrible to look at and detrimental to health. After this careful persuasion, he tempts Trajan further by describing his project as 'tam magno quam necessario' (lines 8-9) — the implication is found in other letters too (e.g. X. 37), that great reigns demand great projects.

Is it with a sardonic smile that Trajan starts his reply, 'It is commonsense, my dear friend Pliny, . . . '? His matter of fact tone suggests that he has seen through Pliny's elaborate selling of the scheme.

Students may find interesting an incongruity revealed about Greek living standards. The Greeks could make Amastris beautiful, but it took the practical Romans to cover an open sewer. At their best the two civilisations became complementary to each other.

Questions

1 What are these two letters about?
2 Did Pliny know a sewer when he saw one?
3 What two reasons does he give for this work being necessary?
4 What does his letter gain by leading up to the point so slowly?
    Is the effect persuasive?
5 Does he think the word 'magno' will put Trajan off?
6 How does Trajan's tone contrast with Pliny's?

X.106, 107

The letters

Pliny sends the Emperor the petition of P. Accius Aquila for a grant of
citizenship to his daughter. Trajan indicates by his rescript that he has
made her a Roman citizen.

Other examples of letters from Pliny in Bithynia concerning individuals and their families: X. 85*; X. 86a*; X. 86b*; X. 87*; X. 94.

Notes: X. 106

1 **domine:** see note on X. 31, line 1.
1-2 **cohortis sextae equestris:** there were no legions in Bithynia. The only
    troops were two cohorts of auxiliaries under Pliny's command and a
    coastal force commanded by an equestrian *praefectus orae Ponticae*
    (X. 21*; S-W, 588-9). It is curious to find the cohort mentioned here
    with only a number and no name. But S-W, 716 refers to a Greek
    inscription giving what is evidently the same title.
2 **ut mitterem tibi libellum:** these words certainly do not suggest the idea
    that soldiers had to go through their provincial governor when petition-
    ing the emperor. It was common practice for people in the empire to
    petition the emperor, often in person (see Fergus Millar, 'Emperors at
    work', 9ff., for instance). But Aquila, rather than seek the necessary
    leave of absence and make the long and expensive journey to Rome,
    chose instead to ask Pliny to send his *libellus* to Trajan via the govern-
    ment postal service. Pliny speeds the *libellus* on its way by writing the
    present covering letter to Trajan, playing on his close ties with his army.
5 **humanitatemque praestare:** Pliny ends with one of Cicero's favourite
    rhythmical *clausulae* ($-\cup--\cup$): and indeed the beautiful archi-
    tectured single sentence of this letter seems to echo Cicero.

Notes: X. 107

3 **libellum rescriptum, quem illi redderes:** Trajan has evidently granted
    Aquila's request by placing a rescript at the bottom of his *libellus*.
    Pliny is to pass this on to Aquila.
    redderes: subjunctive because this is a purpose clause. The main verb

'misi' we naturally translate as 'I have sent' (primary sequence) but Trajan intends it as historic sequence — 'I sent' — in the Roman convention of letter-writing. Hence the necessity of the imperfect subjunctive.

Exploration

Why was Aquila's daughter not a Roman citizen already? It is true that Aquila was only a centurion in an auxiliary unit, and most centurions in the auxiliaries were *peregrini*, not acquiring Roman citizenship till the end of their service. However, Aquila, to judge by his three names, was already a Roman citizen. So his daughter should have been a Roman citizen too, unless Aquila had married a non-citizen or his daughter was illegitimate: either of these may be the explanation.

An auxiliary soldier received the Roman citizenship for himself, his wife and children at the end of his army service. By sending this petition to the Emperor, Aquila was seeking to anticipate the grant he would receive for his daughter on his discharge. Perhaps she was about to marry a citizen and was anxious to become a citizen herself, so that her children would be born *cives Romani*.

The common practice of petitioning an emperor suggests that the emperors were remarkably accessible to those subjects who tried to reach them. So, a litigant approached Vespasian on a journey, while one of his mules was being reshod (Suetonius, *Divus Vespasianus*, 23); and a woman called after Hadrian, when he refused to stop on a journey to listen to her, 'Then stop being emperor' (Dio Cassius, LXIX. 6. 3). Students could profitably read these and other accounts as a basis for a discussion about the accessibility of the emperors. Are modern rulers less accessible, and if so, why?

Such a discussion may shed light on a surprising aspect of Roman government. The emperors tended to preserve the *status quo*, unless prodded into action by individuals or communities. The practice of approaching the emperors with requests indicates that the central government was not unsympathetic to pressure from below but that it was unwilling to take the initiative itself. This unwillingness to act *sua sponte* may have had more than one cause. Perhaps there just was not the machinery of government to make more than a minimum of changes. Perhaps the autocracy of the emperors had killed the initiative of the senatorial class, who could not adapt to the role of civil-service planners. Perhaps the Roman character, so sympathetic to tradition and so happy at incorporating the best of other civilisations into its own culture, was governmentally unadventurous. At all events, this exchange of letters shows clearly enough that emperors were prepared to act favourably on the requests of determined individuals. The forceful and the articulate could get their way, as the centurion Aquila discovered (or already knew).

Questions

1 What has Pliny sent to the Emperor along with his own letter?
2 What does Accius Aquila want?
3 How is the fact of being a soldier to his advantage, according to Pliny?
4 Does Trajan grant the request? Do you think the reply was dictated by him personally?
5 Do you think Aquila might be surprised at Trajan's answer?
6 Why was Pliny ready to help him?

X. II6, II7

The letters

Pliny is concerned at the numbers of guests invited to certain functions. Trajan shares his concern but reminds Pliny — rather curtly — that he is in Bithynia to use his judgement on just such matters.

Notes: X. 116

2 **vel opus publicum dedicant:** cf. IV. 1, where Pliny is planning to go to Tifernum-on-Tiber to attend the dedication of a temple which he has built at his own expense and has decided to celebrate the occasion with a public feast. In these two letters there is no mention of feasts (though public banquets did occur in Asian provinces) but only of a cash distribution.

4-5 **quod an celebrandum ... scribas:** 'I ask that you write whether you think such celebration should take place and to what extent.' A difficulty here is the word-order: the indirect question 'quod an ... putes' is dependent on 'scribas' which in its turn is dependent on 'rogo', with the word 'ut' omitted.

5 **sicut:** this word is taken up by 'ita' two lines later: 'for myself, although I think ... I nonetheless fear that ... '

6 **sollemnibus:** this word, which Trajan also uses (rescript, line 3) denotes something formal or ceremonial.

8 **modum excedere:** it is not clear whether this is a general phrase ('going too far') or whether there was a definite numerical limit laid down by Pliny's *mandata* (see next note).

8-9 **in speciem ... videantur:** both Pliny here and Trajan in the rescript use a Greek word which denotes a distribution of dole or largess; and both imply that it was something prohibited. The Greek word is evidently familiar to Trajan: perhaps it has occurred in his *mandata* to Pliny.

Notes: X. 117

3 **viritim:** this word is really superfluous, as 'singulos' already contains the meaning.
  ad sollemnes sportulas: Trajan sees these distributions as taking on a

pseudo-official air, in contrast to the normal *sportula* or hand-out by patrons to clients. He may be fearing that this is an encroachment on his, or his governor's, prerogative.

4-6   **sed ego . . . profutura:** 'but I chose you with your good judgement with a view to this very thing, that you should in person be in control of regulating the customs of that province of yours and should decide upon such things as would contribute to the lasting tranquillity of that province.' The two difficulties of this sentence are (1) the idiomatic 'prudentiam tuam elegi', where abstract is substituted for personal, and (2) the almost clumsy gerundive phrase in Form C (dative) depending on 'moderareris'. The sentence as a whole has a grandiloquent, official flavour, in contrast to the opening sentence and indeed to Trajan's usual terse style. Might these closing words be an almost exact quotation from the preamble to Pliny's *mandata*?

Exploration

The interest of this exchange of letters centres round two points: (1) the exact nature of Pliny's worry; and (2) the tone of Trajan's rescript.

As to the first point, both Trajan and Pliny imply that any *species* διανομῆς is contrary to the law: gatherings must avoid falling *in speciem* διανομῆς. Since διανομή presumably has its normal sense of 'distribution of money' and since only one or two denarii apiece are being distributed (X. 116, line 4), neither Pliny nor the Emperor can have feared bribery and an infringement of the *Lex ambitus*. However, both men may have feared that even small distributions of money to those in public positions, for example to the whole of the local senate (X. 116, lines 2-3), could be misconstrued by some people. Certainly Trajan in his rescript indicates that money must not be distributed to corporate groups ('per corpora': X. 117, line 2) but only to individuals. Is this why he uses the pleonasm 'viritim singulos', to emphasise the point (X. 117, line 3)?

Pliny however is also concerned that the number of those meeting may seem also to exceed the limit ('modum excedere et . . . ': X. 116, line 8). He must have had at the back of his mind the fear that such gatherings could be construed as an illicit assembly (his edict had banned *hetaeriae* in Bithynia: X. 96 and note). On the other hand, he sees justification for such private and semi-private ceremonies and writes to obtain from the Emperor an assurance that nobody is breaking the law. It is precisely because they are private and semi-private occasions that a doubt has arisen in his mind.

As to the second point — the tone of the rescript — Trajan is evidently a little impatient with Pliny's letter. He implicitly allows these ceremonies to continue but spells it out that numbers must be kept in bounds and payments made only to people as a result of personal acquaintance.

The second sentence of the rescript tells Pliny a little curtly that he was appointed to Bithynia to exercise his own good judgement. This rebuke and that in X. 82* are the standard evidence produced by those

who wish to prove their theory that Pliny was an indecisive, fussy
governor who referred even the smallest points of administration back
to Trajan. Such a thesis is very hard to maintain (see Introduction, III:
Pliny's *Letters*, pp. 15-16). Pliny kept the Emperor informed about his
work in Bithynia, as a special envoy would no doubt be expected to do; he
took pains to convince Trajan of the benefit and necessity of the public
works he was undertaking (X. 41/2; 61, for instance); in two cases at
least he sought to change an existing ruling (X. 31; X. 33*); and, as in
the case of Juliopolis (X. 77), he sought to help the provincials he was
governing. Here in X. 116 he may have been unnecessarily timid. But
Trajan's reply, with the civil service flavour of the second sentence (see
note on X. 117, lines 4-6), may be no more than the Emperor telling his
secretary to give Pliny the standard 'Please make your own decisions'
sentence reserved for governors who could have saved themselves a
letter and the Emperor a rescript. Perhaps in any case Pliny's letter
caught Trajan on a bad day.

Students with a reasonable knowledge of the letters of Book X
should be able to discuss the fairness of Trajan's rescript not just as an
answer to X. 116 but to Pliny's attitude generally.

Questions

1 What practice is Pliny worried about?
2 Why should it be suspect or even illegal?
3 What is the literal meaning of Trajan's phrase 'per corpora'?
4 Is Trajan giving Pliny a polite rebuke? Does he answer Pliny's
   question?
5 Does the style of Trajan's rescript suggest that it was not
   actually worded by him?
6 Need Pliny have written to the Emperor in the first place?

APPENDIX: INSCRIPTIONS

1. Pliny's career

CIL V. 5262; *ILS* 2927; Smallwood (2) 230; S-W, 732; *CLC* Unit III, Stage 31, p. 18. This inscription was evidently at Comum, Pliny's home-town.
See this *Handbook, passim.*

C. Plinius L. f. Ouf. Caecilius [Secundus cos.] augur legat. pro pr.
provinciae Pon[ti et Bithyniae] consulari potesta[t.] in eam provinciam
e[x s.c. missus ab] imp. Caesar. Nerva Traiano Aug. German[ico Dacico
p.p.] curator alvei Ti[b]eris et riparum e[t cloacar. urb.] praef. aerari
Satu[r]ni praef. aerari mil[it. pr. trib. pl.] quaestor imp. sevir equitum
[Romanorum] trib. milit. leg. [III] Gallica[e Xvir stli]tib. iudicand.
ther[mas ex HS - - -] adiectis in ornatum HS $\overline{CCC}$ [- - - et eo amp]lius
in tutela[m] HS $\overline{CC}$ t.f.i. [item in alimenta] libertor. suorum homin.
C HS $\overline{|XVIII|}$ $\overline{LXVI}$ DCLXVI rei [p. legavit quorum in]crement. postea
ad epulum [pl]eb. urban. voluit pertin[ere - - - item vivu]s dedit in
aliment. pueror. et puellar. pleb. urban. HS [$\overline{D}$ item bybliothecam et] in
tutelam bybliothecae HS $\overline{C}$.

> Gaius Plinius Caecilius Secundus, son of Lucius, of the
> Oufentinian tribe, consul, augur, *legatus pro praetore* of the
> province of Pontus and Bithynia being sent into that province
> with consular power by *senatus consultum* by the Emperor
> Caesar Nerva Trajan Augustus Germanicus Dacicus, *pater
> patriae, curator* of the channel of the river Tiber and of its
> banks and of the sewers of Rome, prefect of the Treasury of
> Saturn, prefect of the military treasury, praetor, tribune of the
> people, quaestor of the Emperor, member of the six-man
> commission for the Roman knights, tribune of the soldiers of

the Third Gallica legion, member of the ten-man commission *stlitibus iudicandis,* left by will public baths at a cost of . . . sesterces, with an additional 300,000 sesterces for equipping them, with a further 200,000 sesterces for their upkeep. He also bequeathed to the state 1,866,666 sesterces for the support of one hundred of his freedmen, and it was his wish that the interest of this should subsequently be spent on a banquet for the people of the city. Likewise during his life he gave 500,000 sesterces for the maintenance of boys and girls of the city, as well as 100,000 sesterces for a library and its upkeep.

For an explanation of the horizontal and vertical bars used in the sums of money, see X. 37.

An inscription (*CIL* V. 5667, quoted in *CLC Handbook* to Unit IV, p. 44, slide no. 26), which was found at Fecchio, a small village near Como, and had been set up to Pliny by the people of Vercellae, records *inter alia* that he was priest of the deified Titus, perhaps at Comum.

2. Cornutus Tertullus' career

ILS 1024; MW 321; Smallwood (2), 209.
See Introduction, pp. 5 and 8; VII. 21.
C. Iulio P. f. Hor - - - Cornuto Tertul[lo] cos. proconsuli provinci[ae Asiae] proconsuli provinciae Narbo[nensis] legato pro praetore divi Traiani [Parthici] provinciae Ponti et Bith[yniae] eiusdem legato pro pr[aetore] provinciae Aquitani[ae] c[e]nsu[um] accipiendorum cu[ra]to[ri viae] Aemiliae praefecto aerari Sa[tu]r[ni] legato pro praetore provinc[iae] Cretae et Cyrenarum a[dl]e[cto] inter praetorios a divis Ves[pasiano] et Tito censoribus aedili Ce[riali] quaestori urbano ex testamento C. Iulius Plan(i)cius Varus Cornutus - - -

> To Gaius Iulius Cornutus Tertullus, son of Publius, . . . , consul, proconsul of the province of Asia, proconsul of the province of (Gallia) Narbonensis, *legatus pro praetore* of the deified Trajan Parthicus of the province of Pontus and Bithynia, *legatus pro praetore* of the same Emperor of the province of Aquitania for the conduct of the census, *curator* of the Via Aemilia, prefect of the Treasury of Saturn, *legatus pro praetore* of the province of Crete and Cyrene, adlected into the praetorians' rank by the deified Vespasian and the deified Titus when they were censors, *aedilis Cerialis, quaestor urbanus.* By the terms of his will Gaius Iulius Plancius Varus Cornutus . . .

A valuable example of a successful senatorial career that consisted entirely of civilian appointments.

3. L. Calpurnius Fabatus' career

ILS 2721.
See IV. 1; VII. 16.
[L.] Calpurnius L. f. Ouf. Fabatus VIvir. IIIIvir. i. d. praef. fabr.
trib. iterum leg. XXI Rapac. [pr] aef. cohortis VII Lusitan. [et] nation.
Gaetulicar. sex quae sunt in Numidia [f]lam. divi Aug. patr. munic.
t. f. i.

> Lucius Calpurnius Fabatus, son of Lucius, of the Oufentinian
> tribe, member of a six-man commission, member of the four-
> man commission *iuribus dicundis, praefectus fabrum,* tribune a
> second time of the Twenty-First Rapax legion, prefect of the
> Seventh Lusitanian cohort and of the six Gaetulican nations
> which are in Numidia, priest of the deified Augustus, patron
> of his town. By the terms of his will he ordered this to be
> put up.

('trib. iterum' at the start of the second line of the inscription: Dessau
commented that Mommsen thought this the probable, but not certain
reading.) See notes on IV. 1 for L. Calpurnius Fabatus' career. The importance of this inscription lies partly in the fact that it shows Fabatus'
equestrian career coming prematurely to an end, presumably through
his involvement in the charges against L. Junius Silanus in Nero's reign.
Fabatus did not die till the middle of Trajan's reign (X. 120*, see *CLC*
Unit IV 'Bithynia' and *Handbook*, p. 81).

INDEX

Index